Bayesian Analysis with Stata

John Thompson
University of Leicester

A Stata Press Publication
StataCorp LP
College Station, Texas

Published by Stata Press, 4905 Lakeway Drive, College Station, Texas 77845
Typeset in LaTeX 2ε
Printed in the United States of America
10 9 8 7 6 5 4 3 2 1

ISBN-10: 1-59718-141-2
ISBN-13: 978-1-59718-141-9

Library of Congress Control Number: 2014930482

Dedication

This book is dedicated to Terese and Joanne in recognition of their love and support during the writing.

Contents

Figures

Tables

Preface

Bayesian analysis has become increasingly popular in recent years and proved useful in many areas, including economics, medicine, science, and the social sciences. Much of this growth has been fueled not by philosophical concerns about the nature of statistical analysis or by a genuine desire to use prior information but by very practical considerations; the fact is that many complex models are much easier to handle within a Bayesian framework.

The chief limiting factor in adopting Bayesian methods is the availability of software. Thanks to programs such as Stata, most frequentist statistical methods are easy to implement, but Bayesian analyses require their own specialist software. Following theoretical developments in the 1980s, a major step forward in Bayesian computing came about with the creation of the BUGS project, which provided flexible, free software for fitting Bayesian models. The original BUGS program was eventually replaced by WinBUGS, which in turn has given way to an open-source version called OpenBUGS. Unfortunately, while these programs will fit a model, they are not particularly user friendly, and they do not offer the many facilities for data handling, exploratory statistics, and graphics that Stata users take for granted. Consequently, even investigators familiar with WinBUGS tend to do most of their work in traditional statistics packages and only switch to WinBUGS to fit the model.

Thompson, Palmer, and Moreno (2006) describe a suite of Stata ado-files, all beginning with the prefix `wb`, that help Stata users integrate WinBUGS into their analyses. The idea is that users should be able to store their data in Stata and analyze them in the usual way. Then when they want to fit a Bayesian model, they should be able to prepare it, send it to WinBUGS, and read back the results without ever leaving Stata. The results provided by WinBUGS or OpenBUGS consist of a very large file of simulations, so other programs are provided to help the user read those results into Stata, check the analysis, and summarize the findings. The commands that operate on the results of the WinBUGS run work equally well with Markov chain Monte Carlo (MCMC) simulations produced by any other software, so they are not restricted to use with output from WinBUGS.

Since 2006, OpenBUGS has become available, the original `wb` ado-files have been extended by adding extra options, and further programs have been added to the collection. New and old versions of the commands can be distinguished between those required to run WinBUGS or OpenBUGS and those used to investigate the resulting MCMC simulations; the former have been collected together as the new commands beginning with the letters `wbs`, and the latter now form the new commands beginning with the letters mcmc. Thus the updated version of wbrun that runs WinBUGS from within Stata is called wbsrun, and the updated version of wbtrace, the program for drawing a trace or history plot, is called mcmctrace.

In this book, I describe the updated commands and introduce other programs for running Bayesian analyses that do not need WinBUGS or OpenBUGS. Chapter 3 presents a set of new programs collected under the names beginning with the letters mhs for running Metropolis–Hastings samplers, and chapter 4 introduces programs beginning with the letters gbs for running Gibbs samplers. These programs can create MCMC simulations and fit Bayesian models without using WinBUGS. For multiparameter models, it is convenient to have a housekeeping program that cycles through the parameters and stores the MCMC updates. This job is performed by the program mcmcrun, which oversees the fitting process by calling any user-specified combination of the mhs and gbs samplers or even the user's own samplers.

Creating MCMC simulations independently of WinBUGS serves two purposes: first, it can be used as a teaching aid to demonstrate what happens when a Bayesian model is fit; second, it offers an alternative practical method by which users can tackle real problems. The only limitation of using Stata on its own is the time taken by simulation-based Bayesian model-fitting algorithms. While the samplers with command names beginning mhs and gbs work well on small- or moderately sized problems, efficient programming is essential when there are more than about 10 parameters in the model or when the dataset is very large. This means either using Stata's matrix language, Mata, or exporting the problem to WinBUGS; these two approaches are broadly equivalent in terms of speed and are introduced in chapters 7 and 8. Users who are only interested in using WinBUGS or OpenBUGS can skip chapters 2, 3, 4, 7, and 12, though in doing so, they will miss out on the explanation of how MCMC samplers work.

The author's blog can be found at http://staffblogs.le.ac.uk/bayeswithstata/. It is dedicated to the discussion of the use of Stata for Bayesian analysis and to describing future developments of the ado-files introduced in this book.

Downloading the user-written commands

The user-written commands discussed in this book can be downloaded from within Stata using the following commands:

```
. net from http://www.stata-press.com/data/bas/
. net install bas
```

To download the do-file to install the user-written Mata code, type

```
. net get bas
```

and then run the mcmclibrarymata.do file.

Acknowledgments

Particular thanks are due to the people at StataCorp who supported me during the writing of this book, particularly Bill Rising, who made many useful suggestions mostly concerning the syntax of the Stata code used for the examples.

1 The problem of priors

1.1 Case study 1: An early phase vaccine trial

A researcher tests a new vaccine in an early phase clinical trial by giving it to 10 healthy volunteers, 4 women and 6 men. The researcher is interested in the safety of the vaccine rather than its efficacy, so amongst the many recorded measurements is whether the site of the vaccination becomes tender. In the trial, five of the volunteers report tenderness, one woman and four men. Using these data, the researcher wishes to estimate how many people would experience tenderness if the vaccine were to go forward into an efficacy trial involving 3,000 subjects.

Even this simple problem has more than one plausible solution. If the researcher believes that men and women react differently to the vaccine, then the researcher might argue that if he or she were to recruit 1,500 men and 1,500 women, then $1{,}500 \times 4/6 + 1{,}500 \times 1/4 = 1{,}375$ people who would be expected to experience tenderness. However, if men and women react in the same way, $3{,}000 \times 5/10 = 1{,}500$ people who would be expected to report tenderness.

To decide which estimate to use, the researcher might discuss the problem with experts who have wide experience in vaccine trials or might look back over the records of previous trials of similar vaccines to discover whether men and women reacted differently. Suppose that an expert tells the researcher that there is no biological reason why men and women should react differently and that in the expert's experience of similar vaccines, about 40% of both men and women report tenderness. This seems to tip the balance in favor of the estimate of 1,500, but it also suggests a third estimate based on the expert's judgment: 40% of 3,000 is 1,200. If the researcher is willing to accept the expert's opinion on the difference between men and women, why not also make use of the estimate of the proportion experiencing tenderness? Perhaps the researcher could use the estimate in combination with the data from the early phase trial.

In practice, most statisticians would accept the advice of the expert on whether men and women react differently, but fewer would be willing to use the expert's opinion on the proportion who would report tenderness, even though both types of information affect the final answer. Those statisticians willing to use numerical information that does not come from the study being analyzed are referred to as Bayesians; this book considers the methods of statistical analysis that flow from that willingness and, in particular, how those methods might be implemented in Stata. However, before we get into those technicalities, it is helpful to consider why some people are uneasy about using external numerical information.

The most common objection to the use of an expert's numerical estimate regards subjectivity. The expert's opinion is specific to that individual; had we asked a different expert, we might well have had a different response. When a problem can have different answers depending on which expert we consult, can any answer be relied on? Some would argue that the statistician's job is to summarize the evidence from the current study as objectively as possible. Unfortunately, this argument is undermined by the fact that no analysis is truly objective: subjective judgments are made throughout the planning, conduct, and analysis of any study, and those judgments have an influence on the final answer. For instance, our researcher had to decide whether to adjust for gender and even whether it is sensible to extrapolate from healthy volunteers to the people recruited into an efficacy trial.

Other statisticians argue that although the use of subjective estimates is acceptable in principle, it is impossibly hard to do in practice. To use the expert's opinion, we must decide how much weight to give it relative to the weight that we give to the actual data. In the case of the vaccine trial, this weight will depend on how certain the expert is that 40% of future subjects will experience tenderness. Although 40% is the best guess based on past experience, does the expert think that the percentage for the new vaccine will be something between 35% and 45% or perhaps something between 20% and 60%? Such certainty is difficult to quantify, yet it changes the relative weight given to the expert's opinion compared with the actual data and can greatly influence the final answer. This difficulty is magnified in complex problems in which many factors have to be taken into account because the expert will need to give his or her opinion—with uncertainty—on all of those potentially interrelated factors.

The distinction between Bayesian and non-Bayesian statisticians used to be clear, but in recent years, the barriers between the two traditions have largely disappeared with most statisticians adopting a much more open approach. Experience shows that Bayesian analyses are good for some problems and not so good for others, and unless the expert's opinion is held with great certainty, the actual data will carry more weight in the analysis, and the Bayesian and non-Bayesian answers will be similar.

1.2 Bayesian calculations

Before we consider the benefits of the Bayesian approach, it is helpful to set out a typical statistical analysis in more general terms. To analyze a set of data, y, we must first select a model that describes the trend and natural variation in the data. This model is usually specified in terms of a set of parameters, θ, that allow the model to be tuned to fit the observations. So the model describing the probability of observing a particular set of data can be written as $p(y|\theta)$. Once the general form of the model has been selected, we need a method for using the data to guide us in selecting sensible values for the parameters.

In the case of the vaccine trial, 5 out of 10 subjects reported tenderness, so $y = 5$, and a possible model for the data is the binomial, which depends on a single parameter, θ, representing the probability that an individual reports tenderness. Under this model,

$$p(y|\theta) = \binom{10}{y} \theta^y (1-\theta)^{10-y}$$

In the Bayesian approach, the expert's beliefs about all possible values of the parameter must be specified as a *prior distribution*, $p(\theta)$; in this way, we capture both what we expect to find and our uncertainty. In specifying this prior, we must not be influenced in any way by the data that we are about to analyze, so it is best to specify the prior before seeing the data. The prior could take any form, but for illustration, suppose that the expert believes that the true value of θ could be anything between 0.3 and 0.5 and that within those limits, any value is just as likely as any other. This prior would be represented by a uniform distribution

$$p(\theta) = \begin{cases} 5 & 0.3 < \theta < 0.5 \\ 0 & \text{otherwise} \end{cases}$$

Combining the data and the prior opinion is now a noncontroversial exercise in probability and relies on Bayes' theorem,

$$p(\theta|y) = \frac{p(y|\theta)p(\theta)}{p(y)}$$

Here $p(\theta|y)$ is referred to as the *posterior distribution* of θ because it describes what we believe about θ after combining our prior opinion with the data. $p(y|\theta)$ is the same likelihood that forms the basis of many traditional analyses, and the denominator, $p(y)$, is the probability of the data averaged over all possible values of θ, so $p(y) = \int p(y|\theta)p(\theta)d\theta$.

Figure 1.1 shows our prior and posterior beliefs about θ for the vaccination data. Notice how the observation of 5 out of 10 has shifted our beliefs toward $\theta = 0.5$ and away from $\theta = 0.3$ and also how values totally excluded by the prior will have no posterior support whatever the data might suggest. This plot was created in Stata, but because our interest here is in the principles of Bayesian analysis, the description of the code that was used is postponed until the start of chapter 2.

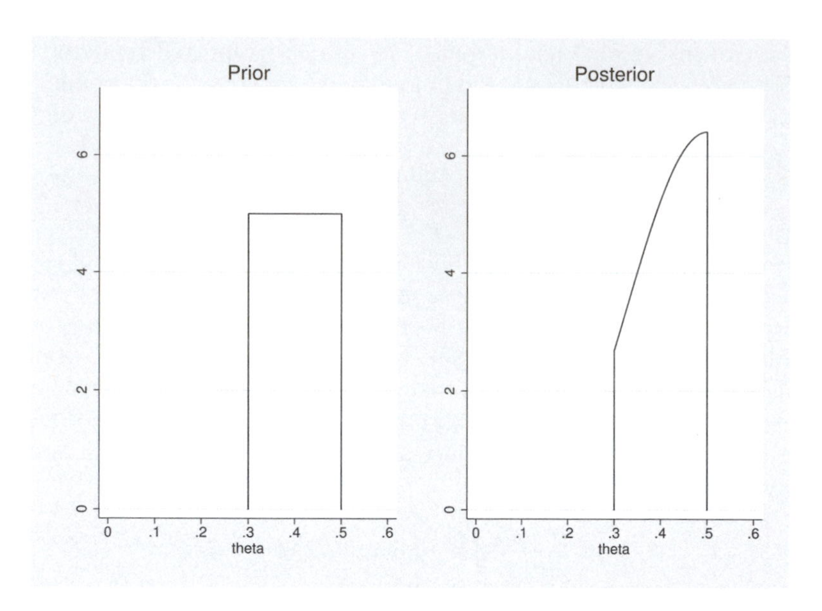

Figure 1.1. Prior and posterior distributions of beliefs about θ

Often the results of an analysis will be used to make predictions about data that will be seen in the future; here we will denote such future data by y^* and its *predictive distribution* by $p(y^*)$ and denote predictions informed by past data, y, by $p(y^*|y)$. Once again the formula for calculating the predictive distribution follows from standard probability theory; it is either

$$p(y^*) = \int p(y^*|\theta)p(\theta)d\theta$$

or, if we have past data available to refine our estimates of the parameters,

$$p(y^*|y) = \int p(y^*|\theta)p(\theta|y)d\theta$$

That is, the predictive distribution based on previous data is derived by averaging the data-generating model over our posterior beliefs about the parameters.

1.3 Benefits of a Bayesian analysis

Once we have calculated the posterior distribution, $p(\theta|y)$, we can make probability statements about θ, for instance, by finding the probability that θ lies within some specified range. Such probabilities have to be interpreted in a similar way to the prior (that is, as subjective expressions of our posterior beliefs about θ), but at least they refer to the actual problem before us. This is in marked contrast to more traditional measures, such as the 95% confidence interval, that rely on repeated sampling for their meaning. Non-Bayesians have to imagine repeated samples of data drawn under the same model so that they can say 95% of such samples would produce confidence intervals that include the true value of θ. Of course, that statement tells us nothing about the properties of our particular confidence interval. The Bayesian analysis is more directly relevant to our study but at the expense of making our statements conditional on both the model and the prior.

Modern simulation methods for calculating posterior distributions have proved so powerful that they have completely turned the tables on more traditional approaches to statistics. Previously, Bayesian statisticians could tackle only relatively simple problems in which the mathematics was tractable or the integrals could be evaluated numerically, while non-Bayesians could tackle many more problems before encountering this type of constraint. With simulation methods, Bayesians can now fit almost any model that they choose, while non-Bayesians still run into computational problems when models contain large numbers of random effects or awkward hierarchical structures. This new freedom has made Bayesian analysis much more attractive and is probably the driving force behind the wider acceptance of the approach.

Now that Bayesian statisticians can choose more or less any model that they think describes the data, the quality of the analyses that they produce ought to have improved. Unfortunately, this new freedom has sometimes led to the creation of needlessly complex models that are almost impossible to verify and that have so many potentially interacting key assumptions that a thorough sensitivity analysis is ruled out. Simplicity is still a virtue, even in Bayesian analysis.

1.4 Selecting a good prior

Even experienced researchers analyzing simple problems find it difficult to specify their prior distribution. Experts unfamiliar with Bayesian analysis find it particularly difficult, and considerable effort has gone into devising methods of eliciting good prior information from individuals and groups. Such methods will always have their limitations, so just as statisticians are used to thinking of their models as being useful approximations rather than as true data-generating mechanisms, researchers should also think of their priors as acceptable approximations.

When the results of an analysis are intended for use by a small number of individuals, then those people can agree among themselves about the form of the prior distribution and so obtain an acceptable Bayesian analysis. However, when research is intended for general consumption, as in the case of a scientific study reported in an academic journal, then the grounds for choosing an appropriate prior are no longer clear. If the authors of the article are allowed to use their own priors, then the possibility exists that some unscrupulous authors will select the priors to skew the results and influence the conclusions. To avoid this, either the authors would have to exert a great deal of effort to explain whether the result is sensitive to the choice of a prior or the authors would have to use some standard or reference prior.

The use of a prior that is not chosen to reflect the views of a particular person or group of people is sometimes referred to as objective Bayesian analysis, although this name is not really appropriate. Usually the standard prior is chosen to have a small influence on the results so that the data dominate. So far the search for a general definition of a minimally informative prior has not been successful; it is common to choose vague priors that are relatively flat over a wide range of values in the hope that they will be noninformative in the sense of having little impact on the final answer.

Early Bayesian analyses were dominated by completely flat priors that extended over the full range of the parameters. When the parameter has an infinite range, a flat prior will not integrate to one and is described as improper. An improper distribution cannot really be said to represent probabilities and at best is an approximation to a uniform distribution over a very large but finite range. Even though improper priors do sometimes lead to proper posterior distributions, their use has fallen out of favor, and they are generally avoided in modern Bayesian analyses.

Before simulation methods were available for calculations in a Bayesian analysis, it was common to choose priors that simplified the mathematics and made the integrals tractable. It became popular to seek out priors that lead to posteriors from the same family. Thus, if we are using a binomial model with parameter θ and choose a beta distribution for the prior, say, beta(a,b), then the posterior distribution of θ will also be a beta distribution but with parameters beta($a + y$,$b + n - y$). Such priors are said to be conjugate. Unfortunately, conjugate priors only exist for a few simple problems, and it is questionable whether the form of the prior should be dictated by mathematical convenience, even if we accept the idea that the prior is only an approximation to our beliefs.

The search for proper noninformative priors is made more complex by the fact that the properties of a prior distribution may be altered by a change of scale. For example, take the problem of setting a prior for a standard deviation. One vague distribution that might be used is a uniform between two limits, say, uniform(0,10). Under this prior, all standard deviations between 0 and 10 are thought equally likely, so the prior probability that the standard deviation is over 6 is 0.4. Now suppose instead that we had decided to work in terms of the variance. We might have opted for a flat distribution over an equivalent range, uniform(0,100); in this case, the probability that the variance is over 36 (standard deviation over 6) is 0.64. A flat distribution for the standard deviation does not give the same results as a flat distribution for the variance.

Even more disturbing than the problem of changing scale is the sensitivity that results can have seemingly small changes in the prior. Once again, take the problem of the prior for a variance. For mathematical convenience, it is common for Bayesians to work on the scale of the precision, which is equal to one over the variance. Because the precision must be positive, a popular vague prior is a gamma distribution with mean one and a large variance. Such a distribution will be relatively flat over a large range of positive values. Unfortunately, as Gelman (2006) has demonstrated, the posterior estimate of the precision can be sensitive to the arbitrary choice of the variance of the gamma prior; thus the gamma prior with large variance has declined in popularity.

When researchers select priors intended to be noninformative, it is tempting to play it safe and extend them over huge ranges, well beyond any value that might realistically occur. However, such impossibly wide priors can create problems when simulation methods are used to investigate the posterior distribution. Very wide priors might cause the simulation to investigate extreme parameter combinations, which could slow down the algorithm or even cause numeric overflow when the posterior is evaluated. Realistically vague priors are generally preferable to impossibly wide priors and have the advantage of making researchers think about their models. Seeking realistic priors will avoid the situation in which, for example, a normal prior with mean 0 and variance 10,000 is routinely used for all parameters that represent means, including perhaps one that represents the average human femur length measured in millimeters.

Three general statements can be made about the selection of noninformative priors: first, automatic choices are potentially dangerous, so it is much better to put some thought into the actual range of plausible values of the parameter; second, vague priors that completely exclude theoretically possible parameter values should be used only after careful thought; and third, whatever choice is made needs to be assessed in a sensitivity analysis to see whether small changes to the prior have a noticeable effect on the results.

1.5 Starting points

Much of the early literature criticizing Bayesian statistics has become outdated because of the enormous progress that has been made in the subject over the last decade or so, and a starting point is the recent article by Gelman (2008). Although the author accepts the Bayesian approach himself, he still manages to convey many of the objections that people have to the use of Bayesian methods, including doubts about subjectivity and the excessive complexity of some Bayesian analyses. The article is followed by a series of equally informative discussions and a rejoinder.

The distinction between objective and subjective Bayesian analysis is well summarized by Berger (2006), who puts the case for objective methods, and Goldstein (2006), who puts the case for subjective methods. These articles are also accompanied by discussion articles and rejoinders. An interview with Jose Bernardo conducted by Irony and Singpurwalla (1997) provides an interesting overview of many different types of priors seen from a subjective standpoint. Natarajan and McCulloch (1998) and Gelman (2006) consider practical problems that can arise when supposedly noninformative priors are used.

Eliciting subjective priors from an expert is a complex problem with its own extensive literature. Good starting places for reading on this topic are the Elicitation of Experts' Probabilities website, http://www.shef.ac.uk/chebs/research/themes/elicitation/, and a systematic review by Johnson et al. (2010).

1.6 Exercises

1. An expert tells you that the prior probability of an event occurring should be described as uniform between 0.3 and 0.4. You conduct an experiment and find that the event occurs only once in 20 occasions.

   a. Sketch the shapes of the prior and posterior distributions.
   b. In light of the data, was the prior reasonable?
   c. Would you be willing to use predictions based on this posterior?
   d. Would it be reasonable to ask the expert to reassess his or her prior?

2. A model is devised for predicting movements in the exchange rate between the euro and the dollar. The intention is to fit the model to data from the last 12 months and then use the model to make predictions.

   a. Is it possible for an expert to give informative priors for the parameters in the model that are not already influenced by the data you hope to analyze?
   b. How might you obtain priors for this problem?

3. An expert gives his or her prior distribution for the average daily sulfur dioxide level at a particular point in the center of town. You measure the sulfur dioxide level daily for a week and use a Bayesian analysis to obtain the posterior distribution. When you show the data and the posterior distribution to the expert, he or she refuses to accept your analysis and draws a completely different distribution, saying that this is what he or she now believes after seeing the data.

   a. Why might your calculation and the expert's sketch disagree?
   b. Which distribution better describes the expert's updated beliefs, your posterior or his sketch?
   c. If you reject the posterior, can you ever again be sure that a posterior distribution describes an updated opinion?
   d. If you reject the sketch, what confidence can you have in the prior that the expert drew?

4. A researcher wishes to estimate the proportions of different types of terrain using aerial photographs of sample areas. He or she agrees to classify the terrain as forest, grassland, or other and, for a particular region, gives prior estimates of the percentages as 60%, 30%, and 10%. When you ask about his or her uncertainty, the researcher says that all prior estimates are accurate to within ±10%; that is, the researcher believes the percentages to lie within the ranges (50%,70%), (20%,40%), and (0%,20%). Further, he or she believes that within those ranges, all percentages are equally likely.

   a. Why do these figures not describe a prior distribution?
   b. How would you get a nonstatistician to describe his or her prior for this problem?

2 Evaluating the posterior

2.1 Introduction

Stata users fitting a generalized linear model with the `glm` command do not need to concern themselves with the model's likelihood or the fitting algorithm, because they are built in to the command. A Bayesian command for fitting a generalized linear model could be programmed in much the same way except that it would also need to incorporate a choice about the form of the prior. Such a restriction would be rather limiting because the program would in effect control the form that the researcher's prior beliefs were allowed to take.

An alternative way of programming is to provide a command that performs Bayesian calculations for any given combination of likelihood and prior but that relies on the user to specify those two functions. There is a strong parallel here with the way that Stata's `ml` command maximizes user-specified likelihoods. Some extra programming is required of the user, although the increased flexibility makes this worthwhile: not only do users have control over the specification of their priors, but the approach encourages them to move away from a reliance on a limited range of standard models.

This chapter describes how to write programs for evaluating any combination of likelihood and prior in a way that can be used in a general-purpose Bayesian algorithm. Although we start with a straightforward, one-parameter problem, the main illustrations involve the use of the programs to calculate conditional posterior distributions in multiparameter models. The more general problem of using the programs to investigate the whole multiparameter posterior distribution is covered in subsequent chapters.

2.2 Case study 1: The vaccine trial revisited

The model for the vaccine trial that was described in chapter 1 has just one parameter, so it is easy to analyze that trial by evaluating the Bayesian integrals using Stata's `integ` command. This particular case study used a binomial model for an observation of 5 out of 10 with a prior on the probability, θ, that was uniform between 0.3 and 0.5. The following code will calculate the prior, likelihood, and posterior.

begin: vaccine trial

```
range theta 0 1 1000
generate Prior = 5*inrange(theta,0.3,0.5)
generate LogL = 5*log(theta)+5*log(1-theta)
summarize LogL
generate Likelihood = exp(LogL-r(max))
integ Likelihood theta
replace Likelihood = Likelihood/r(integral)
generate Posterior = Likelihood*Prior
integ Posterior theta
replace Posterior = Posterior/r(integral)
twoway (line Prior theta) (line Likelihood theta) ///
   (line Posterior theta), legend(pos(1) ring(0) row(3))
```

end: vaccine trial

Although the use of `integ` does not generalize to models with many parameters, there are still several points in the code worth noting. The `range` command defines a set of 1,000 values for the parameter spread evenly between 0 and 1, and the prior and log likelihood at those points are evaluated using the formulas given in section 1.2. Because constants added to the log likelihood do not affect the end result, the terms that do not depend on theta are omitted from `LogL`. The log likelihood is converted into a likelihood and normalized. The prior and log likelihood are then combined to give a quantity that is in proportion to the posterior, which is also normalized to have an area of one using `integ`. For most problems, it is better to work with the log likelihood rather than the likelihood because the likelihood often involves very small quantities that can lead to a loss of precision. To preserve numerical accuracy, one must add an arbitrary constant to the log likelihood before exponentiating it; in this case, we subtract the maximum log likelihood. The resulting plots are shown in figure 2.1. The prior and posterior represent proper densities, but the scaling of the likelihood is arbitrary, and it was only normalized so that it would be easy to display on the same axes as the other curves.

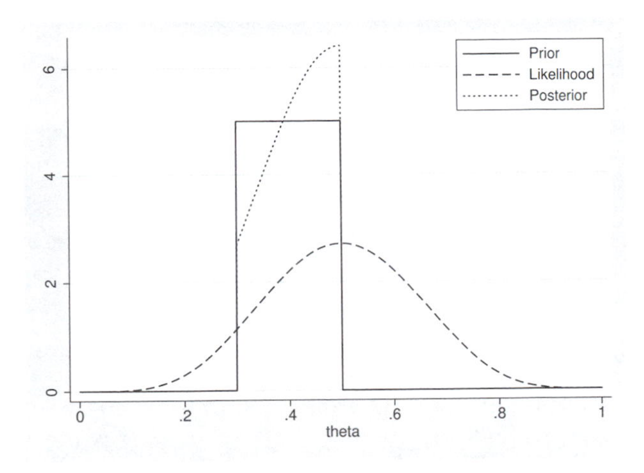

Figure 2.1. Prior, likelihood, and posterior for the vaccine trial

2.3 Marginal and conditional distributions

As already noted, proper probability densities integrate to one, but in a Bayesian analysis, it is easier to work with functions that are proportional to the required density and only to normalize them to have an area of one at the end of the calculations. A function that is proportional to the joint posterior is readily obtained by multiplying together the likelihood and prior because from Bayes' theorem, we see

$$p(\theta|y) \propto p(y|\theta)p(\theta)$$

In this formula, θ may be a vector, so the eventual normalization can be nontrivial because it will involve an integral with dimension equal to the number of parameters, and even after normalization, the high-dimensional joint posterior distribution can be very difficult to visualize. Interpretation of the posterior would be much easier if it were possible to view the separate distributions of the individual parameters; such a simplification would best be achieved through the calculation of either marginal or conditional posterior distributions.

Suppose that the joint posterior distribution of two sets of parameters takes the form $p(\theta_1, \theta_2|y)$. A conditional posterior distribution such as $p(\theta_1|\theta_2 = t, y)$ makes statements about one of the sets of parameters, in this case, θ_1, conditional on some specified value, t, of the other. When θ_1 and θ_2 are scalars, we can visualize the joint posterior as a surface with axes θ_1 and θ_2 so that the conditional distribution is obtained as a slice through the hill along the line $\theta_2 = t$. Such a slice will enclose an area of varying size depending on where the cut is made; thus, to convert the slice into a proper distribution, one needs to normalize it.

$$p(\theta_1|\theta_2 = t, y) = \frac{p(\theta_1, \theta_2 = t|y)}{\int p(\theta_1, \theta_2 = t|y)d\theta_1} \propto p(\theta_1, \theta_2 = t|y)$$

This equation shows that a quantity that is proportional to the conditional posterior distribution can be obtained by plugging the conditioning values into any function that is proportional to the joint posterior. This implies that we can insert the conditioning values into the product of the likelihood and prior and then normalize the resulting low-dimensional function.

The marginal distribution of a parameter, for instance, $p(\theta_1|y)$, is more difficult to evaluate because it averages our beliefs about one parameter over all values of the others. Typically, this involves another multidimensional integral,

$$p(\theta_1|y) = \int p(\theta_1, \theta_2|y) d\theta_2$$

In many situations, the marginal distribution will be a better description of our beliefs about θ_1 than any single conditional distribution because it is not tied to particular values of the other parameters. However, it is only a summary; in effect, the marginal distribution is a weighted average of all the possible conditional distributions. If our beliefs about θ_1 depend strongly on what we know about θ_2, then the marginal distribution may not convey the whole story.

2.4 Case study 2: Blood pressure and age

The website http://archive.ics.uci.edu/ml/datasets/ contains several datasets intended for use in machine-learning experiments. One of these datasets, given on the site under the heading "Heart Disease", comes from an international collaboration formed to identify predictors of coronary disease. Centers in three countries collaborated to collect many potentially predictive variables on patients suspected of coronary heart disease, as subsequently described by Detrano et al. (1989). In our analysis, we will concentrate not on the prediction of patients likely to have heart disease but on the much simpler problem of modeling the relationship between age and blood pressure. We will use the Swiss component of the study, which was conducted by Dr. Steinbrunn and Dr. Pfisterer at the University Hospitals in Basel and Zurich.

In the Swiss dataset, there are 123 subjects, 121 of whom had resting systolic blood pressures taken at referral. Representing the systolic blood pressure (SBP) measurements on the 121 patients with y and regressing them on the patients' ages, x, using a linear model with normal errors gives

$$y_i \sim N(\alpha + \beta x_i, \sigma) \quad i = 1, \ldots, 121$$

This model requires priors on α, β, and σ, and these must be specified without reference to the data. Thus, if we want to make the priors informative, we must either seek the opinion of an expert or find other data to guide our choice. In this case study, obvious sources of information are the other centers in the collaboration; to help picture this external information, figure 2.2 shows the corresponding Hungarian data together with their linear regression fit by least squares. The slope is about 0.55, the intercept at age 0 is about 105, and the residual standard deviation is about 17. However, in setting our priors for the Swiss data, we need to allow not only for the uncertainty in the Hungarian estimates but also for the differences between Switzerland and Hungary. For instance, the referral patterns of suspected heart disease patients might be quite different in the two countries.

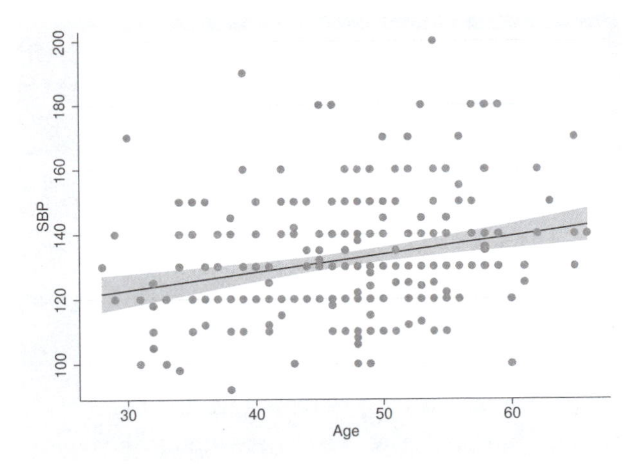

Figure 2.2. Systolic blood pressure (SBP) against age in the Hungarian study

The eventual choice of priors is inevitably subjective, but realistic choices for the Swiss data are

$$\alpha \sim N(105, 10)$$
$$\beta \sim N(0.55, 0.3)$$
$$\sigma \sim U(5, 35)$$

Whenever possible, it is a good idea to run a sensitivity analysis, changing these priors to see what impact they have on the final answer. However, this must not be confused with a search for a better set of priors, and whatever the sensitivity analysis might show, we will have to live with the consequences of our initial choice.

The appendix gives the algebraic forms of several commonly used distributions, but for users who wish to avoid the algebra, there is a downloadable author-written program called `logdensity` that calculates the value of the log density for many standard distributions. This program can be incorporated into user-written Stata programs for evaluating the product of likelihood and prior. However, as we will see shortly, the use of this program is not always efficient, and a little algebra can lead to improvements in the code and faster execution.

To avoid numerical problems resulting from attempts to calculate with very small numbers, we will work with the log of the product of likelihood and prior. Thus the following Stata program could be used to calculate the unnormalized log posterior for the SBP model given any combination of α, β, and σ.

begin: logpost

```
program logpost
    args logp b

    local alpha = `b´[1,1]
    local beta  = `b´[1,2]
    local sigma = `b´[1,3]
    scalar `logp´ = 0
    logdensity normal  `logp´  sbp    `alpha´+`beta´*age `sigma´
    logdensity normal  `logp´ `alpha´ 105 10
    logdensity normal  `logp´ `beta´  0.55 0.3
    logdensity uniform `logp´ `sigma´ 5 35
end
```

end: logpost

When we fit other models, the code for `logpost` will always have this broad structure, so other programs can be written with only basic knowledge of Stata by treating this example as a template.

This program calculates the log of the likelihood times the prior and stores the result in a scalar. It requires the user to supply two arguments, `logp`, the name of the scalar that will store the result, and `b`, a row vector containing the values of the parameters `alpha`, `beta`, and `sigma`. To make the program easier to read, we unpack the parameters from the row vector and store them as locals before using `logdensity` to perform the calculations. This program adds the log density onto the current value of `logp`; thus it is important to set the scalar to zero before `logdensity` is first called. In this example, independence is assumed between the observations and between the parameters in the prior, so all the individual terms can be summed.

`logdensity` accepts arguments corresponding to the name of the distribution, the name of the scalar to which the calculated value is to be added, the data or parameter that follows that distribution, and the parameters of the distribution. `logdensity` checks the arguments, and if the third one is a variable, it is treated as data, and the log density is summed over its values. So here the first call to `logdensity` sums the log likelihood over the data variable, `sbp`, and the other three calls calculate the log priors at a single point.

To illustrate the use of the program, we will calculate the conditional distribution of the slope, β, when $\alpha = 100$ and $\sigma = 20$. In the code below, the program `logpost` is used to calculate the value of the log posterior over a range of values of β running from 0.4 to 0.8 in steps of 0.001. The calculated values are stored in the file `temp.dta`, and the posterior is found by exponentiating and normalizing.

begin: analysis code

```
use swiss.dta, clear
drop if missing(sbp,age)
matrix b = (100,.,20)
tempname pf
postfile `pf´ beta logp using temp.dta, replace
foreach beta of numlist 0.4(0.001)0.8 {
   matrix b[1,2] = `beta´
   logpost logf b
   post `pf´ (`beta´) (logf)
}
postclose `pf´

use temp.dta, clear
line logp beta, name(logp,replace) ytitle(Log-posterior)
summarize logp
generate p = exp(logp - r(max))
integ p beta
replace p = p/r(integral)
line p beta, name(posterior,replace) ytitle(Posterior)
```

end: analysis code

The two plots produced by this code are shown as figure 2.3. Conditional on our beliefs about the other parameters, the value of β is almost certainly in the range 0.45 to 0.65. Marginal statements about β would be closer to the usual regression coefficients of SBP on age and would in most situations be more relevant, but they require two-dimensional integration over α and σ. This is best tackled by the simulation methods described in the following chapters.

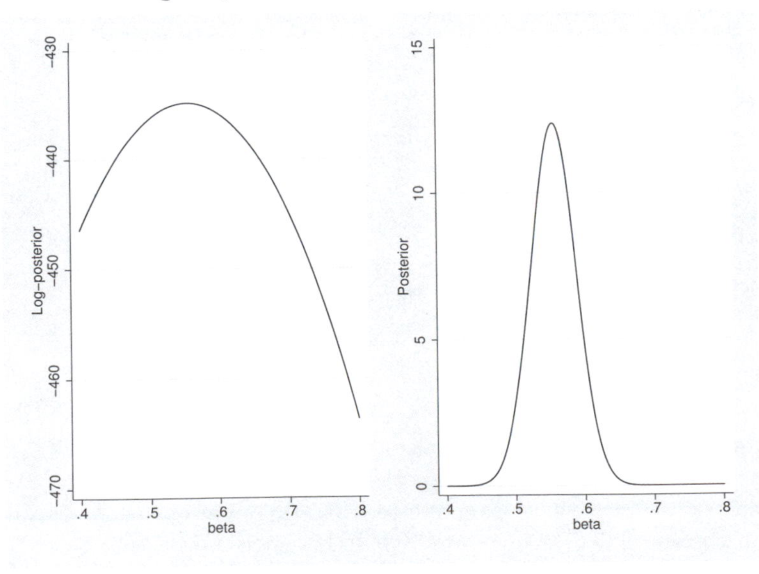

Figure 2.3. Conditional log(likelihood.prior) and conditional posterior distribution for the Swiss SBP regression

The program **logpost** calls **logdensity** and so avoids the need for any algebra. However, for this model, the algebraic form of the log posterior is quite simple, and using the formulas given in the appendix, we can write it as the sum of the component log densities. Recall that y is the SBP, and x is the age:

$$\sum_{i=1}^{121}\left\{-\log(\sigma)-\frac{1}{2}\log(2\pi)-\frac{(y_i-\alpha-\beta x_i)^2}{2\sigma^2}\right\}-\log(10)-\frac{1}{2}\log(2\pi)-\frac{1}{2}\frac{(\alpha-105)^2}{10^2}$$
$$-\log(0.3)-\frac{1}{2}\log(2\pi)-\frac{1}{2}\frac{(\beta-0.55)^2}{0.3^2}-\log(30)-\log(C)$$
$$-\infty<\alpha,\beta<\infty \quad 5<\sigma<35$$

Here C is the normalizing constant that involves a three-dimensional integral. Because the log posterior is only required to within an additive constant, all the terms that do not vary with the three parameters can be ignored, so the expression simplifies to the following:

$$-121\log(\sigma)-\sum_{i=1}^{121}\left\{\frac{(y_i-\alpha-\beta x_i)^2}{2\sigma^2}\right\}-\frac{(\alpha-105)^2}{200}-\frac{(\beta-0.55)^2}{0.18}$$

Indeed, when we calculate the conditional distribution of β when $\alpha=100$ and $\sigma=20$, the only terms that matter are those involving β,

$$-\sum_{i=1}^{121}\left\{\frac{(y_i-100-\beta x_i)^2}{800}\right\}-\frac{(\beta-0.55)^2}{0.18}$$

This simpler expression can be programmed as follows:

begin: faster **logpost**

```
program logpost
    args logp b

    local beta  = `b´[1,2]
    tempvar z
    generate `z´ = -(sbp-100-`beta´*age)^2/800
    quietly summarize `z´, meanonly
    scalar `logp´ = r(sum) - (`beta´-0.55)^2/0.18
end
```

end: faster **logpost**

This reprogramming produces equivalent results, but it can make a noticeable difference in the time taken to evaluate the function. Timing 10,000 evaluations on a desktop computer, the original version of **logpost** took 0.00137s per evaluation, while the function based on the simplified algebra took 0.00021s. Such a difference is irrelevant when evaluating the log posterior once, but in a typical Bayesian analysis, the log posterior might need to be evaluated hundreds of thousands of times. A million evaluations of this log posterior would take just over 22.8 minutes with the original version of **logpost** and about 3.5 minutes with the more efficient code. These considerations can become very important in more complex problems, and it will sometimes even be necessary to program critical parts of a Bayesian analysis using Mata or as a C plugin.

Finally, notice the flexibility of this form of programming. Suppose that there was reason to suspect that some people in the sample will have extreme blood pressures. Instead of a normal errors regression, it makes more sense to use a distribution with longer tails; a t distribution with 4 degrees of freedom is a popular choice in these circumstances. To implement such an analysis, we simply need to change the call to `logdensity` in the original version of `logpost` to the following:

```
logdensity t `logp´ sbp `alpha´+`beta´*age `sigma´ 4
```

Of course, because the model has changed, we might want to reassess our prior beliefs about the parameters.

This flexibility means that we are no longer restricted to analyzing models that other people have specified and programmed; it is possible to use any combination of distributions that is judged suitable for the data.

2.5 Case study 2: BP and age continued

As part of the Swiss study, whether the subjects had chest pain on referral was recorded: 4 of their patients had typical angina, 4 had atypical angina, 17 had chest pain not due to angina, and the rest were free of pain. Suppose that we want to estimate the rate of increase in blood pressure with age in the group with typical angina. We could of course tackle this problem in exactly the same way we did in the previous section except that there would be only four observations and the estimate would not be very reliable. An alternative approach would be to argue that the group with typical angina is unlikely to be different from the other three groups in terms of its regression equation; thus it should be possible to use the regressions for the other groups to help inform the estimate for the angina patients. If this argument seems reasonable, then the only remaining question is how much weight to give to the data on the other groups in comparison with the weight for the group that actually interests us. One solution to this problem would be to create a hierarchical model in which the intercept and slope for each category of patients are drawn from a higher-level distribution, usually assumed to be normal. So the model becomes

$$
\begin{aligned}
y_{ij} &\sim N(\alpha_j + \beta_j x_{ij}, \sigma_j) \quad i = 1, \ldots, n_j \;\; j = 1, 2, 3, 4 \\
\alpha_j &\sim N(\mu_a, \sigma_a) \\
\beta_j &\sim N(\mu_b, \sigma_b) \\
\sigma_j &\sim U(a_s, b_s)
\end{aligned}
$$

where j denotes the category of chest pain, and i denotes the individuals within that category. This type of hierarchical structure is much more commonly used in Bayesian statistics than in more traditional analyses partly because such models are easy to handle using the simulation methods described in the following chapters and perhaps partly because Bayesians are more temperamentally open to using information that does not relate directly to the factor under study.

Under this model, the only terms that involve β_1 are the contribution to the likelihood of the patients in category 1 and higher-level parameters μ_b and σ_b. So when we evaluate the conditional distribution of the slope in the angina patients, `logpost` could be simplified by only including those terms. Here the Stata variable `pain` contains the coded pain categories.

begin: `logpost`

```
program logpost
    args logp b

    local alpha1 = `b´[1,1]
    local beta1  = `b´[1,2]
    local sigma1 = `b´[1,3]
    local mub    = `b´[1,4]
    local sigmab = `b´[1,5]
    scalar `logp´ = 0
    logdensity normal `logp´ sbp `alpha1´+`beta1´*age `sigma1´ if pain==1
    logdensity normal `logp´ `beta1´ `mub´ `sigmab´
end
```

end: `logpost`

After normalization, this would give exactly the same conditional distribution for β_1 as a program for the full posterior distribution over all 18 parameters but with much less computation.

Figure 2.4 shows the conditional posterior distribution of the age slope in angina patients when only analyzing the angina patients and under the hierarchical model. In this case, we have conditioned on a hierarchical distribution for the slopes that has a mean, μ_b, of 0.85 and a standard deviation, σ_b, of 0.1. Marginal posteriors that average over what we believe about μ_b and σ_b can be found using simulation methods described in subsequent chapters (in particular, see exercise 2 in chapter 4).

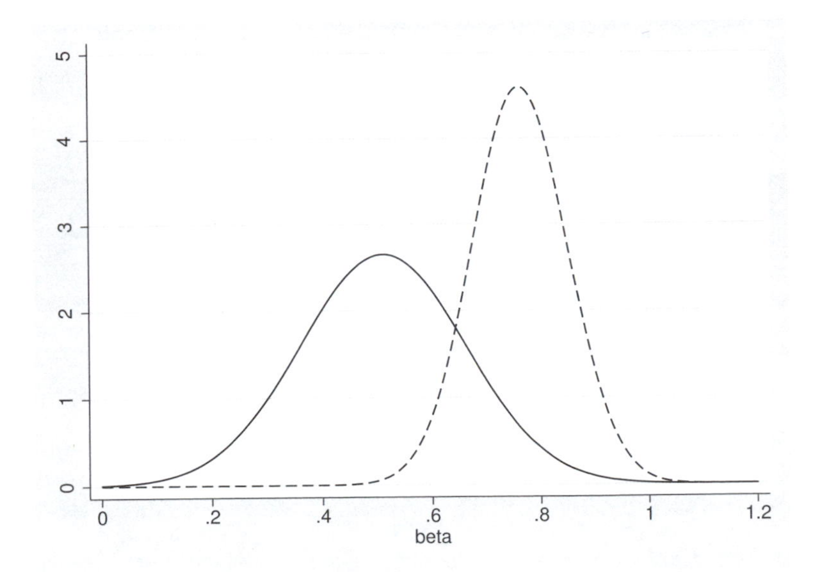

Figure 2.4. Posterior conditional distribution for the Swiss SBP regression in angina patients based on the angina subjects alone (solid) and borrowing strength from the other categories of patient (dash)

2.6 General log posteriors

More complex problems may require a more general program for the log posterior, and it is possible to generalize the structure in several ways. First, in multiparameter problems, it might be important to identify the parameter currently being investigated. This is done by adding a third argument, `ipar`, as in the following example:

begin: `logpost` with parameter number

```
program logpost
    args logp b ipar

    local alpha = `b´[1,1]
    local beta  = `b´[1,2]
    local sigma = `b´[1,3]
    scalar `logp´ = 0
    logdensity normal `logp´ sbp "`alpha´ + `beta´*age" `sigma´
    if `ipar´ == 1      logdensity normal  `logp´ `alpha´ 105 10
    else if `ipar´ == 2 logdensity normal  `logp´ `beta´  0.55 0.3
    else if `ipar´ == 3 logdensity uniform `logp´ `sigma´ 5 35
end
```

end: `logpost` with parameter number

Here the prior for α is only included in the calculation when α is the focus of our attention. When the conditional distributions of β or σ are being investigated, α will be constant. Thus there is no need to include it in the calculations. For illustration, the formula for the mean in the likelihood has been written with spaces between some of the items. For this to be passed to **logdensity** as a single item, it is necessary to place it in quotes.

Sometimes, as in section 2.5, it is important to limit the data used in calculations by using **if** or **in**, and with discrete distributions, it might be convenient to add frequency weights. To achieve this without coding a specific condition into the program, one must write the code in a more general form by using Stata's **syntax** command in place of **args**. For instance, the previous program might be written as follows:

begin: logpost with if and in

```
program logpost
    syntax anything [fweight] [if] [in]

    marksample touse
    tokenize `"`anything´"´
    local logp "`1´"
    local b "`2´"
    local ipar = `3´

    local alpha = `b´[1,1]
    local beta  = `b´[1,2]
    local sigma = `b´[1,3]
    scalar `logp´ = 0
    logdensity normal  `logp´ sbp `alpha´+`beta´*age `sigma´ ///
        [`weight´`exp´] if `touse´
    if `ipar´ == 1      logdensity normal  `logp´ `alpha´ 105 10
    else if `ipar´ == 2 logdensity normal  `logp´ `beta´  0.55 0.3
    else if `ipar´ == 3 logdensity uniform `logp´ `sigma´ 5 35
end
```

end: logpost with if and in

Thus **logpost** could be called with a command such as this:

```
logpost logp b 2 if pain == 1
```

In this example, the parameters are extracted from **anything** by tokenizing the string. Provided that this general structure is followed, all the downloadable programs referred to in the subsequent chapters will be able to communicate with the program.

Finally, a note about numerical precision. When one uses Stata's **ml** command for maximum likelihood analysis, it is important to perform all calculations using double precision so that accurate numerical derivatives can be found. In general, log-posterior calculations are not so sensitive to numerical accuracy, but there are some circumstances in which precision is important. Typically, the calculation of the log posterior will involve summing over the terms in the log likelihood using Stata's **summarize** command. **summarize** returns the sum and other statistics over all the nonmissing values. Consequently, if in your calculation of the log likelihood, the loss of precision causes some of the terms to become missing (perhaps because of division by zero), then those terms will be dropped from the sum. The program will continue, apparently without error, but with the wrong log posterior.

Missing values are only likely to be generated in the calculation of the log likelihood when your priors place probability on very extreme sets of parameter values or when in a search of the parameter space, you start at a very extreme position. For most models, this will not be a problem, but when it is, the calculations will appear to place high-posterior probability on extreme and unlikely combinations of parameter values.

There are two strategies that can be used to address the problem. The first option is to use double precision in the hope that the values will no longer be missing; the second option is to incorporate code to check whether any of the terms are missing before using the sum. If the problem persists, even with double precision, the algorithms that search the parameter space can often be steered away from such extreme values by checking for missing values and, if they are found, replacing the log likelihood by an extremely large negative number. This will make the posterior probability appear negligible at that point. However, should the problem of missing terms in the log likelihood occur frequently, it probably means your model is poorly specified.

2.7 Adding distributions to logdensity

`logdensity.ado` includes code for most of the commonly used distributions, but others could be added by adopting the following template. The parameters passed to `logdensity` must be in order: the name of the distribution; a scalar to contain the evaluated log density; the data in the form of a scalar, a matrix, or a variable containing a set of observations; and the parameters as scalars, matrices, or variables. The `confirm` command is used to see whether the log density is to be evaluated for a scalar, summed over a variable, or, in the case of a multivariate distribution such as the Wishart, calculated for a matrix. When adding further distributions, one must include code for each data type, although some may not be relevant and should lead to an error message. Weights and `if` and `in` qualifiers are only applied to data stored in a variable.

begin: template for `logdensity`

```
program logdensity
    syntax anything [fweight] [if] [in]
    version 12
    local distribution "`1´"
    local logp "`2´"
    capture confirm matrix `3´
    if !_rc {
       quietly logdensity_`distribution´ "M" `0´
    }
    else {
       capture confirm variable `3´
       if !_rc {
           marksample touse
           markout `touse´ `3´
           tempname lnf
           quietly logdensity_`distribution´ "V"  `touse´ `lnf´ `0´
           quietly summarize `lnf´ [`weight´`exp´] if `touse´, meanonly
           scalar `logp´ = scalar(`logp´) + r(sum)
           local n = r(N)
           quietly summarize `touse´ [`weight´`exp´] if `touse´, meanonly
           if r(N) != `n´ scalar `logp´ = .
       }
       else {
           tempname slogp
           scalar `slogp´ = scalar(`logp´)
           quietly logdensity_`distribution´ "S" `0´
           scalar `logp´ = `slogp´ + scalar(`logp´)
       }
    }
end

program logdensity_normal
    local type "`1´"

    if "`type´" == "V" {
        tempvar u
        local touse `2´
        local lnf `3´
        local y (`6´)
        local mu (`7´)
        local sd (`8´)
        generate `u´ = (`y´-`mu´)/`sd´ if `touse´
        generate `lnf´ = -log(`sd´) - 0.5*`u´*`u´ if `touse´
    }
    else if "`type´" == "S" {
        tempname y mu sd u
        local logp `3´
        scalar `y´ = `4´
        scalar `mu´ = `5´
        scalar `sd´ = `6´
        scalar `u´ = (`y´-`mu´) /`sd´
        scalar `logp´ = - log(`sd´) - 0.5*`u´*`u´
    }
    else {
        display as err "No normal logdensity for data in a matrix"
        exit 198
    }
end
```

end: template for `logdensity`

The code in `logdensity` itself should not be changed; instead the user should simply write a program with a name such as `logdensity_`*xxx*, where *xxx* is the name of the distribution. Within this function, the scalar `type` is supplied by the main routine and denotes whether this particular `logdensity` is to be calculated for a variable (`V`), a scalar (`S`), or a matrix (`M`). The user's program does not need to be stored with the standard distributions in the `logdensity.ado` file, although it might be stored there for convenience if the same distribution is to be used many times. As an example of what is required, the code is shown for the normal distribution. Notice that the program does not include the $\sqrt{2\pi}$ term, because it is not needed when calculating the log of a quantity that is proportional to the posterior.

2.8 Changing parameterization

Before finishing our consideration of the calculation of the posterior, we need to note what happens if the parameterization of a model is changed. There are two main reasons why it might be desirable to change parameterization. One possibility is that having derived the posterior, we wish to make statements about some transformed parameter. For instance, in a regression problem, the joint posterior distribution of regression coefficients β_1 and β_2 might have been calculated when it would be interesting to make statements about the ratio β_2/β_1. The second possibility is that priors are specified using one parameterization, but the integrals are easier to evaluate on some transformed scale. So for purely mathematical or numerical convenience, it could be better to reparameterize.

The role of the parameter, θ, in the likelihood, $p(y|\theta)$, where it conditions probability statements about the data, is quite different from its role in the prior, $p(\theta)$, or the posterior, $p(\theta|y)$, where it is the subject of the probability statements. If we wish to transform the parameterization from θ to ϕ, where $\theta = g(\phi)$, then in the likelihood, a direct substitution produces the correct result, $p\{y|g(\phi)\}$, but in the densities, the situation is slightly more complex.

For a continuous prior distribution, the probability associated with a value in the range $(\theta, \theta + \delta\theta)$ is approximately $p(\theta)\delta\theta$. On the transformed scale, this becomes $p\{g(\phi)\}\delta\theta$, but to complete the transformation, we must also convert the small increment using $\delta\theta = g'(\phi)\delta\phi$. So the transformed prior takes the form $p\{g(\phi)\}g'(\phi)\delta\phi$ and depends in part on the derivative of the transformation. It is possible that the transformation could reverse the direction of the scale, as when large values of θ are associated with small values of ϕ. This would create negative derivatives and negative increments on the new scale. To avoid this complication, we use the absolute value of the derivative, and the new increments are treated as positive; that is, we transform to $p\{g(\phi)\}|g'(\phi)|\delta\phi$, so the new density will be $p\{g(\phi)\}|g'(\phi)|$.

In multiparameter problems, there will be increments corresponding to each dimension, and they all need to be transformed. Suppose that the original parameters are $\theta_1, \ldots, \theta_r$ and that they are to be transformed using a set of functions $\theta_i = g_i(\phi_1, \ldots, \phi_r)$. The transformed probability density will have the form

$$p\{g_1(\phi_1, \ldots, \phi_r), \ldots, g_r(\phi_1, \ldots, \phi_r)\}|J|$$

where J is the Jacobian matrix with (i,j)th element $(\partial g_i)/(\partial \phi_j)$, and $|J|$ is its determinant.

The Jacobian complicates any change of parameterization, but there are two instances when the transformations are much simpler. First, discrete distributions give probabilities directly without the need for small increments and so do not require a Jacobian. Second, linear transformations have derivatives that are constant; hence, the Jacobian will also be constant. If all that is needed is a quantity that is proportional to the transformed density, then the constant Jacobian can be ignored.

2.9 Starting points

The best source of information on writing programs in Stata is the [P] *Stata Programming Reference Manual*, but there is also a helpful book by Baum (2009) that gives numerous examples of Stata programs for specific problems. The approach to programming the log posterior is somewhat similar to the way that the `ml` command requires that the user program the log likelihood; that approach is described in the book on `ml` by Gould, Pitblado, and Poi (2010), although, of course, programs created for the Bayesian and likelihood analyses are not interchangeable.

The Bayesian approach to hierarchical models is well described in many books, including Gelman et al. (2004). The introductory article by Greenland (2000) provides an overview that includes both Bayesian and non-Bayesian explanations of the principles behind hierarchical or multilevel models.

Change of parameterization is covered in most textbooks on statistics, such as Hogg, McKean, and Craig (2013), and on numerous webpages that can be located by using a search engine.

2.10 Exercises

1. The Health Insurance Plan (HIP) of Greater New York breast cancer trial as described at http://www.stat.ucla.edu/cases/ was initiated in 1963 and recruited 62,000 women aged 40–64 who were members of the HIP of Greater New York. The women were randomized to either be offered breast cancer screening or not, and 10,800 of those offered screening refused the offer. Mortality during the first 5 years is summarized in the table below.

Table 2.1. Five-year mortality data from the HIP breast cancer screening trial

Group	Number	Number of deaths	
		Breast cancer	Other causes
Offered screening			
Examined	20,200	23	428
Refused	10,800	16	409
Controls	31,000	63	879

a. The total number of deaths from all causes (breast cancer+other) in controls is to be modeled by a Poisson distribution parameterized in terms of θ_0, the death rate per 100,000 person-years. U.S. Census Bureau data on the numbers of deaths from all causes per 100,000 women per year for the United States are given below to help with the specification of your priors. Just over 5% of these deaths were due to breast cancer. Sketch your own prior for θ_0. Choose values for the parameters of a gamma distribution that approximates those prior beliefs.

Table 2.2. Deaths per 100,000 women in one year in the United States

Year	35–44yrs	45–54yrs	55–64yrs
1960	229	527	1106
1970	231	517	1099

b. Write a Stata program with the structure described in this chapter that evaluates the log posterior of θ_0 under the Poisson model and your chosen gamma prior. Use your Stata program to evaluate the log posterior of θ_0 at 1,000 equally spaced points between 0 and 1,000. Convert the calculated values to give values proportional to the posterior, and use Stata's `integ` command to normalize the posterior to have an area of one. Plot the posterior and the prior of θ_0 on the same axes.

c. Repeat the previous steps for the total number of deaths in the 31,000 women offered screening, parameterizing in terms of θ_1, the rate per 100,000 person-years.

d. Was the intervention beneficial?

2. When hospital patients are scanned, they are often asked to take a contrast agent that helps improve the quality of the image. One potential side effect of these agents is that they can temporarily affect the kidneys, a condition known as contrast nephropathy. To counter this side effect, doctors may give patients the drug acetylcysteine. To assess the benefit of this drug, Birck et al. (2003) performed

a meta-analysis of seven trials that had compared acetylcysteine plus hydration with hydration alone in patients with preexisting kidney disease. The results for the seven trials are given in the table below.

Table 2.3. Number of patients with contrast nephropathy/total number of patients

Trial	Treated	Control
1	1/41	9/42
2	8/45	6/40
3	6/92	10/91
4	2/25	13/29
5	10/38	9/41
6	2/60	15/61
7	4/102	12/98

The Stata command `iri` will calculate the relative risk for a single trial and give a 95% confidence interval. The data can then be converted to a log relative-risk (logRR) scale with an approximate standard error. For instance, for trial 1, we could use the following commands:

```
iri 1 9 41 42
display %8.3f log(r(irr)) %8.3f log(r(lb_irr)) %8.3f log(r(ub_irr))
display %8.3f (log(r(ub_irr))-log(r(lb_irr)))/3.92
```

The seven trials will be analyzed using a fixed-effects model for the calculated logRRs,

$$\log \mathrm{RR}_i \sim N(\theta, \sigma_i) \qquad i = 1, \ldots, 7$$

a. Specify a prior for θ that is neutral about whether the treatment will be of benefit to this high-risk patient group.

b. Plot the posterior distributions of θ and of $\mathrm{RR} = \exp(\theta)$ conditional on all σ_j's being equal to their corresponding calculated standard errors.

An alternative analysis could be based on a random-effects model:

$$\log \mathrm{RR}_i \sim N(\mu_i, \sigma_i) \qquad i = 1, \ldots, 7$$
$$\mu_i \sim N(\theta, \phi)$$

c. Under this random-effects model, find the posterior distributions of θ and $\exp(\theta)$ conditional on all σ_j's being equal to their corresponding calculated standard error and $\phi = 0.5$.

3 Metropolis–Hastings

3.1 Introduction

Integration is an essential part of any Bayesian analysis, and for many years, the difficulty of evaluating those integrals prevented the practical use of Bayesian methods, which were restricted to illustrative problems with integrals that could be evaluated analytically or to simple problems with few parameters for which numerical integration was feasible. Then, in the 1980s, statisticians started to experiment with simulation-based methods of integration, and although this work started slowly, it was eventually so successful that it led to an explosion in the use of Bayesian methods. Indeed, for some complex models, fitting is now easier within a Bayesian framework using a Monte Carlo algorithm than it is using the corresponding likelihood maximization.

The idea of simulation is simple enough: to study the characteristics of a probability distribution, $p(\theta)$, one needs only a large representative sample of values $\theta^{(1)}$, $\theta^{(2)}$, $\theta^{(3)}$, ... drawn from that distribution. The characteristics of the sample can then be used to describe the distribution. For instance, the mean of the sample is an estimate of the mean of the distribution, and the proportion of the sample that is positive is an estimate of the probability that θ is positive.

Unfortunately, many of the distributions that arise in Bayesian analyses have such complex forms that drawing the sample of values is not straightforward. A neat solution to this simulation problem is to use a Markov chain, that is, a chain in which each simulated value is randomly drawn dependent on its immediate predecessor. The resulting method will be a Markov chain Monte Carlo (MCMC) algorithm. A long chain will cover the whole probability distribution, even though short subsections of the chain will be correlated and may describe only part of the distribution.

By far, the most widely used of the simulation methods available for Bayesian analysis is the Metropolis–Hastings (MH) algorithm, which works as follows. Suppose that the current value of the chain is $\theta^{(t)}$. MH generates the next value in two stages. First, a possible value, θ', is drawn randomly from a proposal distribution of our own choosing, $q(\theta^{(t)},\theta')$. As the notation suggests, this proposal is allowed to depend on the current value, $\theta^{(t)}$. Second, the proposal is accepted with probability, $\alpha(\theta^{(t)},\theta')$, or it is rejected. Accepted values are added to the chain, $\theta^{(t+1)} = \theta'$, but when the proposal is rejected, the next value in the chain is set equal to its predecessor, $\theta^{(t+1)} = \theta^{(t)}$. The two problems for the statistician are to find the acceptance probability that makes the chain representative of the target distribution and to choose a proposal distribution that makes the algorithm efficient by minimizing the length of chain needed to cover the whole distribution.

Tierney (1994) and, in a less formal but more readable form, Chib and Greenberg (1995) describe how the acceptance probability is derived. The probability, π, that the chain actually moves from $\theta^{(t)}$ to θ' will depend on the proposal distribution and acceptance probability,

$$\pi(\theta^{(t)}, \theta') = q(\theta^{(t)}, \theta')\alpha(\theta^{(t)}, \theta')$$

What is more, any chain can be shown to converge to the density $p(\theta)$ if it has "detailed balance"; that is,

$$p(\theta^{(t)})\pi(\theta^{(t)}, \theta') = p(\theta')\pi(\theta', \theta^{(t)})$$

Combining these equations leads to a relationship between the acceptance probabilities for the moves in opposite directions. One solution to this is to make the acceptance probability the following:

$$\alpha(\theta^{(t)}, \theta') = \min\left\{\frac{p(\theta')\, q(\theta', \theta^{(t)})}{p(\theta^{(t)})\, q(\theta^{(t)}, \theta')}, 1\right\}$$

This choice is known as the MH algorithm after Metropolis et al. (1953), who suggested the basic idea, and Hastings (1970), who generalized it to the form described here. Because this method will be used to explore posterior distributions, it will be most useful in the form

$$\alpha(\theta^{(t)}, \theta') = \min\left\{\frac{p(\theta'|y)\, q(\theta', \theta^{(t)})}{p(\theta^{(t)}|y)\, q(\theta^{(t)}, \theta')}, 1\right\}$$

A convenient property of the acceptance probability is that densities $p()$ and $q()$ occur on both the top and bottom of the expression, so any constants in the formulas for those distributions will cancel. This is particularly important for complex posterior distributions of the form

$$p(\theta|y) = \frac{p(y|\theta)p(\theta)}{\int p(y|\theta)p(\theta)d\theta}$$

because the integral at the bottom of the expression occurs in both $p(\theta'|y)$ and $p(\theta^{(t)}|y)$ and so cancels and does not need to be evaluated. Consequently, the formula for the posterior, $p(\theta|y)$, can be replaced in the acceptance probability by the unnormalized product of the likelihood and prior, giving

$$\alpha(\theta^{(t)}, \theta') = \min\left\{\frac{p(y|\theta')\, p(\theta')\, q(\theta', \theta^{(t)})}{p(y|\theta^{(t)})\, p(\theta^{(t)})\, q(\theta^{(t)}, \theta')}, 1\right\}$$

Another special case arises when the proposal distribution is symmetrical, $q(\theta^{(t)}, \theta') = q(\theta', \theta^{(t)})$, because it will cancel completely from the acceptance probability, leaving

$$\alpha(\theta^{(t)}, \theta') = \min\left\{\frac{p(y|\theta')\, p(\theta')}{p(y|\theta^{(t)})\, p(\theta^{(t)})}, 1\right\}$$

This is sometimes called the Metropolis algorithm and is the form originally described by Metropolis et al. (1953).

Although the choice of proposal distribution is essentially arbitrary, it will have consequences for the speed of convergence of the chain. A proposal distribution that makes suggestions that are frequently rejected produces a chain with large numbers of repeated values that will be very slow to cover the whole of the target distribution; this is usually referred to as poor mixing. Such a poorly designed chain might have to be run for millions of simulations before it covers its target, while a well-chosen proposal distribution for the same problem might converge in a few thousand simulations.

A simple but effective way of generating new proposals is to use a random walk in which $q(\theta^{(t)}, \theta')$ takes the form of a distribution centered on the old value. A normal distribution is a popular choice,

$$\theta' \sim N(\theta^{(t)}, \sigma)$$

with σ chosen by the user to suit the scale of the target distribution. This choice gives the same chance of moving from $\theta^{(t)}$ to θ' as it gives to moves from θ' to $\theta^{(t)}$ and so produces a Metropolis algorithm.

A normal proposal distribution would not work well for a nonnegative parameter such as a variance, because the proposals that it produces might sometimes be negative. A simple alternative is to remove the nonnegative constraint by working with the log of the parameter. This approach can be thought of either as taking the proposal from a log-normal distribution or as working with a transformed parameter. Unfortunately, the transformed proposal distribution is not symmetrical, and the full MH algorithm is needed; however, the theory is still simple, and the algorithm is easy to program. The move can be generated on the log scale by using

$$\log(\theta') \sim N\{\log(\theta^t), \sigma\}$$

On the transformed scale, the acceptance probability must include the Jacobian of the transformation as described in section 2.8. Thus the acceptance probability becomes

$$\alpha(\theta^{(t)}, \theta') = \min\left\{\frac{p(\theta'|y)\ \theta'}{p(\theta^{(t)}|y)\ \theta^{(t)}}, 1\right\}$$

3.2 The MH algorithm in Stata

When one programs the MH algorithm, the values of the posterior distribution that are needed in the acceptance probability often become extremely small. To avoid numerical problems, users should work on a log scale. In this form, the algorithm will

- propose a new parameter value;
- calculate the log of the acceptance probability, $\log(\alpha)$;
- generate a uniform value between 0 and 1, u; and
- accept the proposal if $\log(u) < \log(\alpha)$ or otherwise repeat the previous value.

The code below proposes a value from a normal random walk with $\sigma = 0.5$ and starts with $\theta = 0$. Because the proposal distribution is symmetrical, it cancels in the acceptance probability:

begin: Metropolis

```
matrix theta = (0)
logpost logp theta
matrix newtheta = theta[1,1] + 0.5*rnormal()
logpost newlogp newtheta
scalar logA = newlogp - logp
matrix theta[1,1] = cond(log(runiform())<logA,newtheta[1,1],theta[1,1])
```

end: Metropolis

Here `logpost` is the user's program for calculating the log posterior and has the structure described in chapter 2, while `c=cond(x,a,b)` is a Stata function that puts `c=a` if condition `x` holds and `c=b` otherwise. The parameter is placed in a matrix to make it easier to generalize the code for use in multiparameter problems. This code can be placed in a loop and repeated as many times as necessary.

Given user-written programs `genprop` to generate proposals and `logprop` to calculate the log density of the proposal distribution, the more general MH algorithm for a one-parameter problem could be coded as follows:

begin: MH

```
matrix theta = (0)
logpost logp theta
genprop theta newtheta
logpost newlogp newtheta
logprop logf theta newtheta
logprop newlogf newtheta theta
scalar logA = newlogp + logf - logp - newlogf
matrix theta[1,1] = cond(log(runiform())<logA,newtheta[1,1],theta[1,1])
```

end: MH

3.3 The mhs commands

To save a little on the programming, a set of downloadable author-written commands beginning with the letters `mhs` implements various random-walk MH algorithms. `mhsnorm` uses a normal proposal distribution and is suitable for parameters defined over all real values. `mhslogn` operates on a log scale and is suitable for parameters that must be positive. `mhstrnc` uses a normal proposal on a logistically transformation scale,

$$\log \frac{\theta - l}{u - \theta}$$

which makes it suitable for parameters defined over a finite range (l, u).

The full syntax for each of these commands is described in their help files. It takes the general form:

`mhs`*cmd logpost rowvector parameter* [*if*] [*in*] [*weight*], `sd(#)` [*options*]

The terms following the command are the name of the user's program that calculates the log posterior, the row vector of parameters, and the column number of the parameter to be updated. All three commands require the `sd()` option containing the standard deviation of the proposal distribution, while `mhstrnc` also requires options `lbound()` and `ubound()` containing the lower and upper limits of the range. All the commands allow another option, `logp()`, which can be set equal to the value of the log posterior at the current parameter value if it is known. When `logp()` is not specified or is set equal to a missing value, the value of the log posterior at the current point is calculated. When `logp()` is specified, it reduces the number of calls to *logpost* from two to one and will nearly halve the run time of the program.

Each of the `mhs` commands replaces the parameter in the row vector, `b`, with its update and returns the calculated value of the log posterior at the updated position as `r(logp)`; it also returns a scalar, `r(accept)`, which is set equal to 1 when the proposal is accepted and 0 otherwise. The code below illustrates how `mhsnorm` might be used in a one-parameter problem assuming that the program `logpost` for calculating the log posterior has already been written.

begin: use of `mhsnorm`

```
tempname pf
postfile `pf´ theta using temp.dta, replace
matrix b = (0.5)
forvalues i=1/200 {
   mhsnorm logpost b 1, sd(0.1)
   post `pf´ (b[1,1])
}
postclose `pf´
```

end: use of `mhsnorm`

The single parameter is stored in a 1×1 matrix, `b`, which is initially set to the starting value of 0.5. `mhsnorm` creates random-walk proposals from a normal distribution with standard deviation of 0.1, and the chain is posted to the results file.

3.4 Case study 3: Polyp counts

Xie and Aickin (1997) analyzed data from a phase III colorectal cancer-prevention trial regarding the ability of wheat bran fiber supplement to prevent the recurrence of colorectal polyps. Subjects with polyps were recruited, their polyps were removed, and the subjects were randomized to extra fiber or no extra fiber and followed up with to see whether the polyps recurred. Table 3.1 shows, for the women in the trial, the number of polyps each woman had when she was recruited. So, for example, 251 women entered the trial after having one polyp removed.

Table 3.1. Number of polyps removed from women recruited into the cancer-prevention trial

	Number of polyps removed									
	1	2	3	4	5	6	7	8	9	10
Frequency	251	85	26	17	8	2	0	0	1	0

The Poisson model is sometimes used to model the number of polyps, y, found in a person randomly drawn from the general population. However, in this trial, only subjects with one or more polyps were recruited. Under a Poisson model with rate λ, the proportion of the population with at least one polyp is $1 - \exp(-\lambda)$, so a possible model for the probability of observing y polyps would be the zero-truncated Poisson model:

$$p(y|\lambda) = \frac{\lambda^y e^{-\lambda}}{y!(1 - e^{-\lambda})} \qquad y = 1, 2, \ldots$$

In screening studies of the general population, the prevalence of polyps is typically found to be of the order of 10% to 20%, which corresponds to $0.1 < \lambda < 0.2$. However, the subjects for this trial were drawn from people referred for colonoscopy, and the prevalence might therefore be expected to be considerably higher. A conversation with the surgeons would be needed to obtain an informative prior. In the absence of that, we will use a realistically vague prior for λ in the form of a gamma distribution with shape and scale parameters 3 and 0.3. This prior can be plotted using

```
twoway function y=gammaden(3,0.3,0,x), range(0 3) ///
   xtitle(lambda) yscale(off)
```

and the resulting plot is shown in figure 3.1. We expect lambda to be below 1 but accept the possibility that it might be as high as 2 or 3.

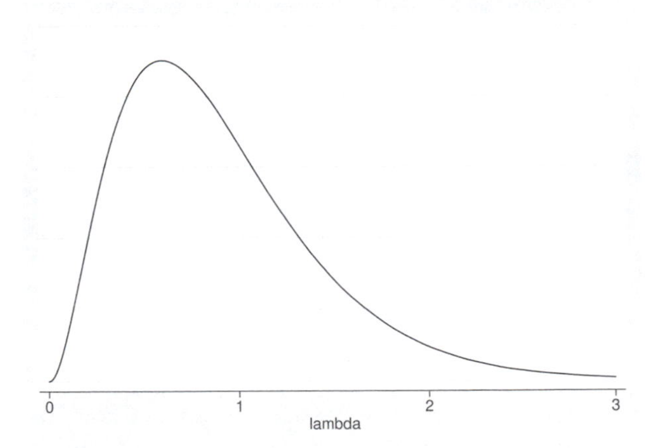

Figure 3.1. Realistically vague prior for the rate of polyps in women

The gamma distribution is widely used in Bayesian statistics, and it is important to realize that there are two commonly used parameterizations: if the shape parameter is a and the scale parameter is b, then what Stata and many other programs call $G(a, b)$ is parameterized as $G(a, 1/b)$ by other software, including WinBUGS. While the parameterization is arbitrary, users should check which version is being used. Here we adopt the Stata parameterization so that $G(a, b)$ has a mean equal to ab.

Assuming that the data have been entered into two Stata variables `y` $= 1, 2, \ldots, 10$ and `f` $= 251, 85, \ldots, 0$, the following code evaluates the log likelihood times prior. The factor 390 arises because there are 390 women in total, and each of their distributions needs to be adjusted for the zero truncation.

begin: **logpost**

```
program logpost
    args logp b

    local lambda = `b´[1,1]
    scalar `logp´ = 0
    logdensity poisson `logp´ y `lambda´ [fw=f]
    scalar `logp´ = `logp´ - 390*log(1-exp(-`lambda´))
    logdensity gamma `logp´ `lambda´ 3 0.3
end
```

end: **logpost**

To implement the MH algorithm, we will use a random walk with a log-normal proposal distribution reflecting the fact that λ must be positive. The code below makes 200 simulations starting with an initial guess of $\lambda = 0.5$ and posts them to the file `temp.dta`. The value of the log posterior is stored after each iteration and supplied back to `mhslogn` using the `logp()` option to reduce the number of log-posterior evaluations.

begin: polyp analysis

```
tempname pf
postfile `pf´ lambda using temp.dta, replace
matrix b = (0.5)
local lp = .
forvalues i=1/200 {
   mhslogn logpost b 1, sd(0.5) logp(`lp´)
   local lp = r(logp)
   post `pf´ (b[1,1])
}
postclose `pf´
use temp.dta, clear
mcmctrace lambda
```

end: polyp analysis

The values of the chain can be plotted using `mcmctrace`, and the resulting plot is shown in figure 3.2. The plot shows the successive values of the chain together with horizontal lines at the median and 95% limits of the sequence. There is an early period during which the chain moves from the initial value of $\lambda = 0.5$ toward the center of the posterior distribution. Then after about 10 iterations, the chain moves backward and forward across the distribution. However, because of the dependence between successive values, it can take many iterations to go from one tail of the distribution to the other. The early section of the chain influenced by the choice of initial value is called the burn-in and needs to be discarded.

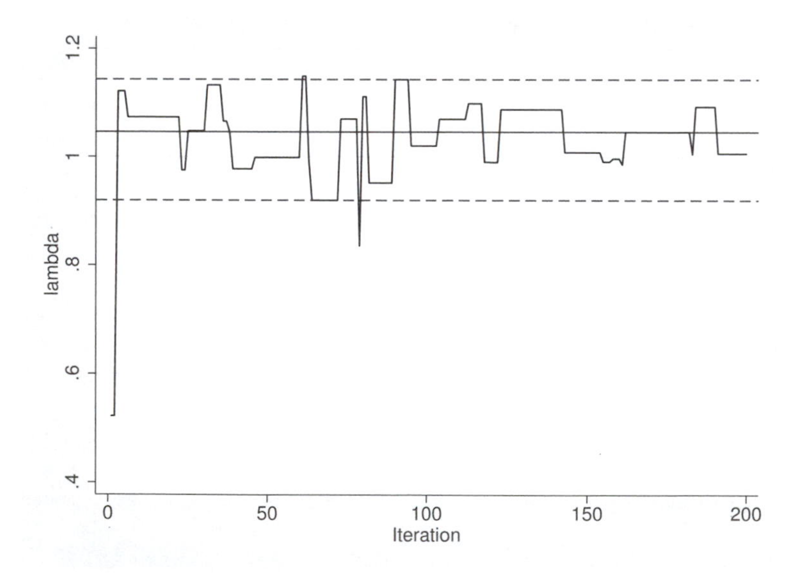

Figure 3.2. Trace plot of the MH chain of 200 simulations for the polyp counts

With a poorly mixing chain, there will be a large correlation between successive simulations, and very little information will be lost if only a sample from the chain is stored. Storing, say, every 10th value is referred to as thinning, and in complex problems with many parameters and perhaps requiring hundreds of thousands of simulations, thinning can greatly reduce the storage requirements without much loss of information.

Figure 3.2 shows a chain that mixes quite poorly because there are runs of rejected proposals that create flat regions in the plot. A smaller standard deviation for the proposal distribution will lead to smaller steps, which would be more likely to be accepted. So to investigate the posterior distribution of λ in more detail, we rerun the code with the proposal standard deviation changed from 0.5 to 0.1, a burn-in of length 500, and a further run of length 10,000 from which every 5th value is stored. The trace plot for a sample of 200 iterations taken from the center of the stored chain, and the trace plot for the entire chain are shown in figure 3.3. The full trace plot shows no obvious drift that would make us concerned that the chain is still influenced by the initial value, and the mixing is improved because the full range of the distribution is covered in a much smaller number of iterations.

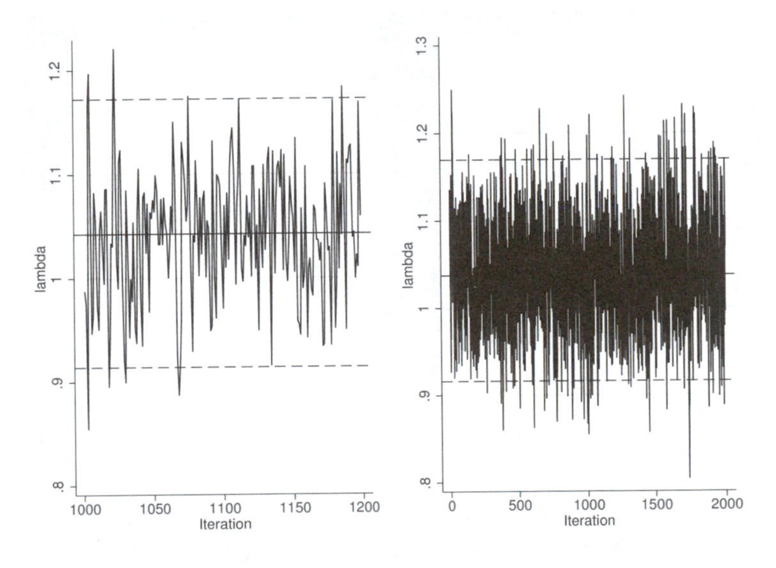

Figure 3.3. Trace plots of a short subsection of the MH chain and of the full stored chain of 2000 simulations

Using Stata's `summarize` command, we find the mean of the simulations to be 1.041. This is similar to the estimate of 1.043 obtained by Xie and Aickin (1997) using maximum likelihood, and the agreement reflects our choice of a rather noninformative prior. Using Stata's `ci` command for confidence intervals would not be valid, because `ci` assumes independent observations, and these results come from a Markov chain. `mcmcstats` produces results that are corrected for this dependence and gives standard deviations, standard errors, and intervals. In this case, the standard error of the posterior mean is 0.0016, which gives an impression of the potential inaccuracy due to sampling and helps in the decision as to whether a longer chain is required. This standard error tells us that the simulated chain has given about two decimal places of accuracy in its estimate of the posterior mean, which is probably sufficient for most practical purposes. This sampling error of the simulations is not to be confused with our uncertainty about the value of λ, which is described by the full range of the posterior distribution, which is between about 0.9 and 1.2 according to figure 3.3.

```
. mcmcstats lambda
----------------------------------------------------------------------------
Parameter           n       mean        sd       sem    median         95% CrI
----------------------------------------------------------------------------
 lambda          2000      1.039     0.064    0.0017      1.037 (   0.916,    1.170 )
----------------------------------------------------------------------------
```

The use of `mcmcstats` is considered in more detail in chapter 6.

3.5 Scaling the proposal distribution

When `mhsnorm`, `mhslogn`, or `mhstrnc` is used to generate the chain, each transforms the scale and then uses a normal random walk centered on the old value with a standard deviation supplied by the user. If the proposal distribution's standard deviation is too large, the moves will be too extreme and will tend to be rejected, so the chain rarely moves. On the other hand, if the standard deviation is too small, the moves will usually be accepted, but it will take many such moves to cover the whole posterior distribution.

To investigate the relationship between the standard deviation of the proposal distribution and the performance of the MCMC algorithm, we reanalyzed the polyp data using `mhslogn` and various standard deviations. The results for runs of 5,000 simulations are shown in table 3.2.

Table 3.2. Standard error of the posterior mean (sem) and percentage of moves rejected for different proposal standard deviations

proposal st dev	sem	percent rejected
0.05	0.0029	25%
0.10	0.0022	43%
0.15	0.0018	57%
0.20	0.0020	67%
0.25	0.0020	70%
0.30	0.0023	76%
0.35	0.0022	80%
0.40	0.0024	81%
0.45	0.0027	83%
0.50	0.0027	85%

First, it is clear that performance, judged by the standard error of the posterior mean (sem), is not very sensitive to the choice of standard deviation for the proposal. Perhaps surprisingly, the best performance is obtained when more than half the proposals are rejected. These results suggest that we should aim for a chain that rejects something between half and two-thirds of the proposals. This finding is consistent with the guidelines suggested by Roberts and Rosenthal (2001). They also showed that for a high-dimensional sampler, an acceptance rate as low as 20% is desirable.

It is not legitimate to change the standard deviation of the proposal distribution dependent on the current performance of the chain because that could affect the shape of the posterior distribution to which the chain converges. However, we can change the proposal standard deviation during the burn-in because these simulations will be discarded. This opens the possibility of tuning the proposal standard deviation during the burn-in to obtain an optimum value, which is then kept fixed during the actual MCMC run.

The commands `mhsnorm`, `mhslogn`, and `mhstrnc` return a value `r(accept)`, which equals 1 if the proposed move was accepted and 0 otherwise. A simple form of tuning is to start with a guess at the appropriate standard deviation, to increase it every time a proposed move is accepted, and to decrease it every time the proposal is rejected. Equivalent steps up and down will produce a chain that tends toward a 50% acceptance rate. Other rates can be created by altering the ratio of the upward and downward movements. Code for such a procedure might take the following form:

begin: adjusting the proposals

```
local s = 0.5
forvalues i=1/1000 {
    mhslogn logpost b 1, sd(`s´)
    if r(accept) == 0 local s = `s´/1.01
    else local s = `s´*1.01
}
```

end: adjusting the proposals

For the polyp data, the log scale was used because λ must be positive. Because λ lies between about 0.9 and 1.2, $\log(\lambda)$ will range from about -0.1 to 0.2. So the optimum proposal standard deviation in table 3.2 is between half and three-fourths of the range of the posterior. This gives us an approximate rule for choosing a starting value for the tuning algorithm. Because the posterior range on a log scale is about 0.3, it would be sensible to start with a proposal standard deviation in the range of 0.15 to 0.20.

3.6 The mcmcrun command

`mcmcrun` is a housekeeping command that does no calculation itself but instead calls samplers such as `mhsnorm`. It provides a convenient way of specifying the samplers, specifying the length of the run, adjusting the proposal standard deviation, and saving the chain to a text file. The full syntax is given in the help file, but briefly, it is `mcmcrun` *logpost rowvector* `using` *filename* [*if*] [*in*] [*weight*], `samplers(`*string*`)` [*options*]

The command name is followed by the name of the program for calculating the log posterior and the name of the row vector containing the parameter estimates. The `using` clause specifies the comma-delimited output file, and the other options are listed in the help file. The only essential option is `samplers()`, which contains the names of the chosen MH algorithms with their options. Assuming that the command `logpost`, described in section 3.4, has already been created and the data have been read, we could fit the zero-truncated Poisson model with these commands:[1]

begin: example of `mcmcrun`

```
matrix b = (0.5)
matrix s = (0.1)
mcmcrun logpost b using temp.csv, samp(mhslogn, sd(s[1,1])) ///
  replace burn(500) up(10000) param(lambda) dots(0) thin(5) adapt
import delimited temp.csv, clear
```

end: example of `mcmcrun`

This code would run the analysis and display the following basic housekeeping information in the Results window:

```
 MCMC using user program logpost
 1 parameters
 Initial values
    0.5000
Samplers
lambda         mhslogn logpost b 1    , sd(s[1 , 1])
Length of Burn-in = 500
Adapting sd during the burn-in

Burn-in completed in      0.390 seconds
Parameter values after burn-in
    1.0641
Run of length 10000 has started
Run saved to temp.csv

Run completed in      7.879 seconds
```

The results can then be inspected; for instance, `matrix list s` displays the adjusted proposal standard deviation used for the updates, and the contents of `temp.csv` describe the posterior distribution.

1. With versions before Stata 13, the `insheet` command can be used in place of `import delimited`.

3.7 Multiparameter models

Nothing in the derivation of the MH algorithm says that θ cannot be a vector of parameters, so the same algorithm could be applied directly to a multiparameter problem. The difficulty would be in suggesting a good proposal distribution for a vector of parameters. When a new vector, θ', is proposed, it would need to be chosen so as to allow for all the interrelationships between the individual parameters; otherwise, there would be a high chance that the proposal would be rejected. The end result would be very slow mixing and an impractical algorithm.

It would be far easier to make proposals if we could treat each parameter separately and just cycle through them; as Tierney (1994) explains, such an algorithm converges to the correct posterior distribution. This very powerful result justifies us in using the MH algorithm to cycle through the parameters one at a time provided that each step is conditioned on the current values of all the other parameters. Indeed, the parameters can be divided into subsets in any convenient way, and the subsets can be updated in any predetermined order.

Suppose that the vector of parameters, θ, is divided into two subvectors, θ_1 and θ_2. The acceptance probability for a proposal, θ_1', would be given by

$$\alpha(\theta_1^{(t)}, \theta_1') = \min\left\{\frac{p(\theta_1', \theta_2^{(t)}|y)\; q(\theta_1', \theta_1^{(t)})}{p(\theta_1^{(t)}, \theta_2^{(t)}|y)\; q(\theta_1^{(t)}, \theta_1')}, 1\right\}$$

Because

$$p(\theta_1|\theta_2, y) = \frac{p(\theta_1, \theta_2|y)}{p(\theta_2|y)}$$

the acceptance probability is unchanged if it is calculated from the full posterior or from the appropriate conditional distributions,

$$\alpha(\theta_1^{(t)}, \theta_1') = \min\left\{\frac{p(\theta_1'|\theta_2^{(t)}, y)\; q(\theta_1', \theta_1^{(t)})}{p(\theta_1^{(t)}|\theta_2^{(t)}, y)\; q(\theta_1^{(t)}, \theta_1')}, 1\right\}$$

Once θ_1 has been updated so that either $\theta_1^{(t+1)} = \theta_1'$ or $\theta_1^{(t+1)} = \theta_1^{(t)}$, we can update θ_2 by creating a proposal, θ_2', and accepting it with probability,

$$\alpha(\theta_2^{(t)}, \theta_2') = \min\left\{\frac{p(\theta_2'|\theta_1^{(t+1)}, y)\; q(\theta_2', \theta_2^{(t)})}{p(\theta_2^{(t)}|\theta_1^{(t+1)}, y)\; q(\theta_2^{(t)}, \theta_2')}, 1\right\}$$

3.8 Case study 3: Polyp counts continued

To illustrate the multiparameter MH algorithm, we will consider an alternative model for the polyp count data. The Poisson model assumes that polyps occur according to a random process with a constant rate, yet one might expect that a person who develops one polyp will be at increased risk of developing others. This would cause polyps to cluster within people who are prone to develop them and violate the assumption of the Poisson model. The result of such a clustering would be overdispersion of the data whereby, the variance is larger than predicted by the Poisson distribution.

An alternative way to model these data is to use a generalized Poisson (GP) distribution. The distribution is not unlike the negative binomial and is mentioned in the book on negative binomial regression in Stata by Hilbe (2011). This distribution can be parameterized in many ways, one of which uses two parameters, μ and α:

$$p(y|\mu,\alpha) = \frac{1}{y!}\left(\frac{\mu}{1+\alpha\mu}\right)^{y}(1+\alpha y)^{y-1}\exp\left\{-\frac{\mu(1+\alpha y)}{1+\alpha\mu}\right\}$$

Under this parameterization, the mean of the GP distribution is μ, and the variance is $\mu(1+\alpha\mu)^2$. Clearly, if $\alpha = 0$, the variance is not inflated, and the distribution is equivalent to a simple Poisson. One needs to take care with the ranges of the parameters in this model: μ must be positive, and α will be positive in the presence of overdispersion. However, α can take negative values, which represent underdispersion, but the allowable range of the negative values depends on μ.

Eliciting informative priors for μ and α would need careful thought, but for illustration, we will adopt a gamma(3,0.3) prior for the mean, μ, and match this with a gamma(0.5,2) prior for α. The prior on α favors small amounts of overdispersion and excludes the possibility of a variance that is less than the mean.

To convert the GP into a zero-truncated distribution, we must scale it by the probability of having at least one polyp, which is

$$p(y>0|\mu,\alpha) = 1-\exp\left(-\frac{\mu}{1+\alpha\mu}\right)$$

There is no prewritten program for the log of the GP distribution, let alone the zero-truncated GP, but it is easy to prepare a program that takes the numbers of polyps, `y` = 1, 2, ..., 10, and frequencies, `f` = 251, 85, ..., 0, and calculates the log of the likelihood times prior. Without making any attempt to make this program general (in the sense of preparing it for use with other datasets), we could code it as follows:

begin: logpost

```
program logpost
    args logp b

    local mu    = `b´[1,1]
    local alpha = `b´[1,2]
    local theta = `mu´/(1+`alpha´*`mu´)
    tempvar ay lnp
    gen `ay´ = 1+ `alpha´*y
    gen `lnp´ = y*log(`theta´)+(y-1)*log(`ay´)-`theta´*`ay´-log(1-exp(-`theta´))
    quietly summarize `lnp´ [fw=f]
    scalar `logp´ = r(sum)
    logdensity gamma `logp´ `mu´    3 0.3
    logdensity gamma `logp´ `alpha´ 0.5 2
end
```

end: logpost

In this code, the constant $y!$ can be omitted from the calculation of the log likelihood because it will cancel in the calculation of the MH acceptance probability. To create an exploratory MCMC run, one might use this code:

begin: two-parameter MCMC

```
matrix b = (1, 0.1)
mcmcrun logpost b using temp.csv, ///
   samp((mhslogn, sd(0.1)) (mhslogn, sd(0.2))) ///
   param(mu alpha) update(2000) replace
import delimited temp.csv, clear
mcmctrace mu alpha, cgopt(row(2))
```

end: two-parameter MCMC

In the resulting trace plot, shown as figure 3.4, mixing is poor, and a careful inspection of the plot shows why: the traces move in unison; when `mu` is high, `alpha` is low, and vice versa.

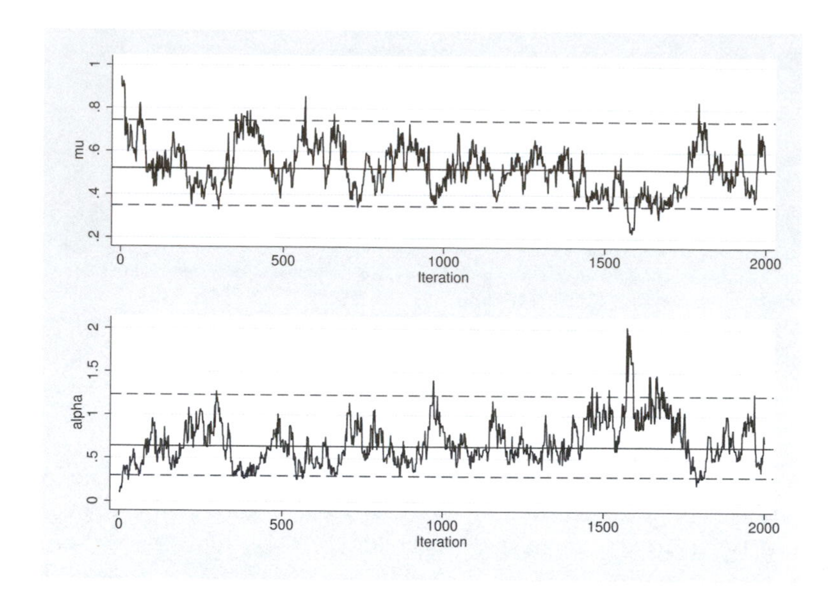

Figure 3.4. Trace plot of the initial chain of 2,000 simulations for the GP model

The correlation is even more obvious from a scatterplot of the simulated values of `mu` against `alpha`, as shown in figure 3.5. This leaves us with two possible explanations for our data: 1) the mean number of polyps per person is around 1, and polyps occur randomly over the population; or 2) the mean is much lower, but a few people have lots of polyps.

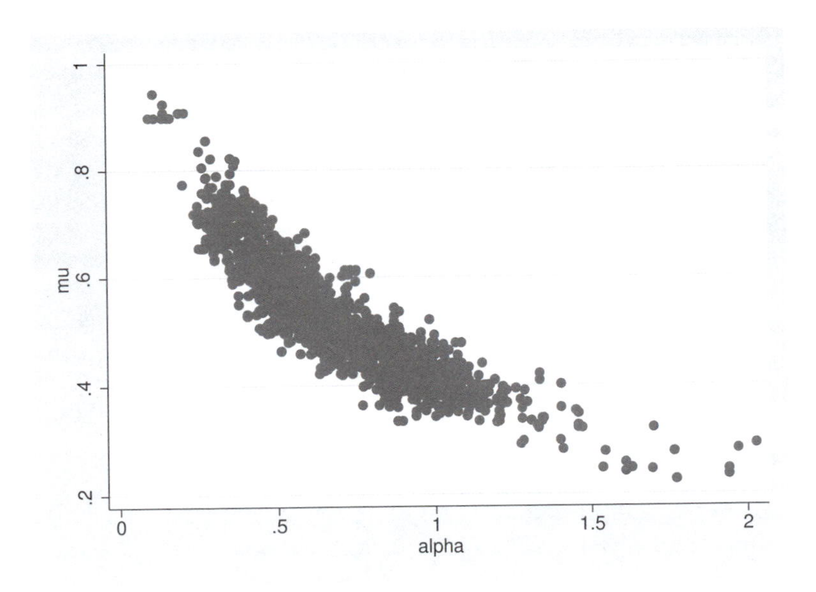

Figure 3.5. Scatterplot of 2,000 simulations of the parameters of the GP model

To see whether the explanation for the poor mixing lies with our choice of proposal standard deviations, we can investigate the frequency with which the proposals were rejected using

```
count if mu == mu[_n-1]
count if alpha == alpha[_n-1]
```

In this case, the proposal for μ was rejected 660 times in 2,000 simulations, and the proposal for α was rejected 770 times. These are perhaps a little low but not dramatically so. The problem here is not so much our proposal distribution as the inherent correlation between the parameters. Techniques for handling correlated parameters will be discussed in the next section. In the meantime, we will adopt a brute-force solution by running a long chain. A burn-in of 1,000 simulations was followed by a run of 100,000 thinned by 5 so that the stored chain had 20,000 values. This produced the following results:

```
. mcmcstats mu alpha
----------------------------------------------------------------------------
Parameter          n      mean       sd       sem    median         95% CrI
----------------------------------------------------------------------------
 mu            20000     0.493    0.128    0.0038     0.494 (   0.247,    0.745 )
 alpha         20000     0.825    0.453    0.0152     0.715 (   0.304,    1.973 )
----------------------------------------------------------------------------
```

Even a run length of 100,000 leaves us with a sem of 0.015 for `alpha`, which translates to a possible sampling error of about ± 0.03 or 4% of the estimate. A run of about a million simulations would be needed to reduce this to 1%.

3.9 Highly correlated parameters

When two or more parameters are highly correlated, the mixing of an MCMC algorithm that updates each parameter in turn can become slow. At each stage, parameters that are highly correlated with the parameter being updated are held fixed, so there will not be much scope for change, and allowable steps will be small. In this situation, there are four main strategies for improving the mixing:

- Reparameterize the problem so that the new parameters are less highly correlated.
- Update the set of correlated parameters as a block using a multivariate proposal distribution.
- Overrelaxation.
- Select a multidimensional direction for the move based on current knowledge about the shape of the posterior.

Overrelaxation is available as an option in WinBUGS and will be considered in chapter 8, although it could also be implemented in a user-written sampler. The last of these options involves running several chains in parallel, so at any stage, many sets of parameters are drawn from different parts of the posterior. Any one of the chains can then be updated using the current information about the posterior that is derived from the other chains. This approach is the subject of much current research under the general title of population-based sampling, but these methods are complex and only worthwhile for extremely difficult posterior distributions. In contrast, the first two techniques in the list are much more likely to be useful and so will be considered in some detail.

3.9.1 Centering

The classic use of reparameterization comes when one of the parameters is regressed on a covariate. For instance, suppose that the parameter μ depends linearly on the covariate x:

$$\mu = \alpha + \beta x$$

Typically, α and β will be highly correlated because increasing the slope, β, will tend to decrease the intercept, α. To illustrate this situation, we used the code below to generate a small dataset in which μ is the mean of a normal distribution and the true values of the parameters are $\alpha = 90$, $\beta = 0.9$, and $\sigma = 25$.

begin: data simulation

```
set seed 826015
set obs 50
generate x = round(40+30*runiform())
generate y = round(90+0.9*x+25*rnormal())
```

end: data simulation

These data were analyzed under a linear regression model with priors chosen to be vague enough that they would not materially influence the results:

$$
\begin{aligned}
y_i &\sim N(\mu_i, \sigma) \\
\mu_i &= \alpha + \beta x_i \\
\alpha &\sim N(100, 100) \\
\beta &\sim N(1, 1) \\
\sigma &\sim U(0, 50)
\end{aligned}
$$

A burn-in of 1,000 simulations followed by a run of 5,000 gave the results shown in figure 3.6. The high correlation between the simulated values for α and β is clear.

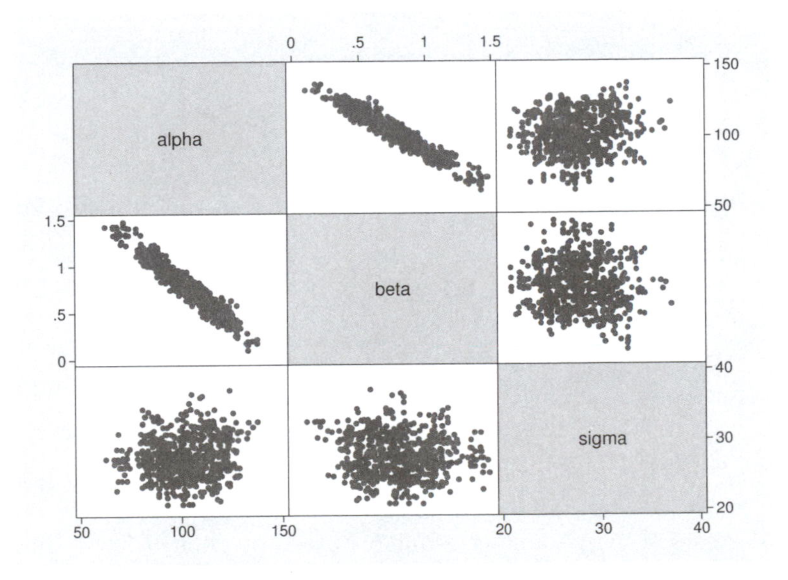

Figure 3.6. Stata scatterplot matrix of the uncentered analysis of the regression problem

A simple reparameterization will greatly reduce the correlation between the simulated values of α and β. Pick a constant, m, close to the mean of the covariate, x, and write the model as follows:

$$\mu = \alpha^* + \beta(x - m)$$

The new parameter, α^*, represents the value of μ when x equals m and will be much less dependent on the slope. Consequently, α^* and β will have low correlation, and a standard MCMC algorithm will mix much better. Once the simulations have been run, it will be possible to transform the simulated values back to the original scale using

$$\alpha^{(t)} = \alpha^{*(t)} - m\beta^{(t)}$$

In this way, we can take advantage of the improved mixing of the second parameterization while still having the estimates for the original parameterization. Of course, one must take care to specify an appropriate prior distribution for α^* when running the reparameterized model. If this is not consistent with the original prior on α, then the two analyses will give slightly different results.

The scatterplot matrix of the centered analysis is shown in figure 3.7. The correlation is visibly reduced, and the mixing of the chain is correspondingly much better.

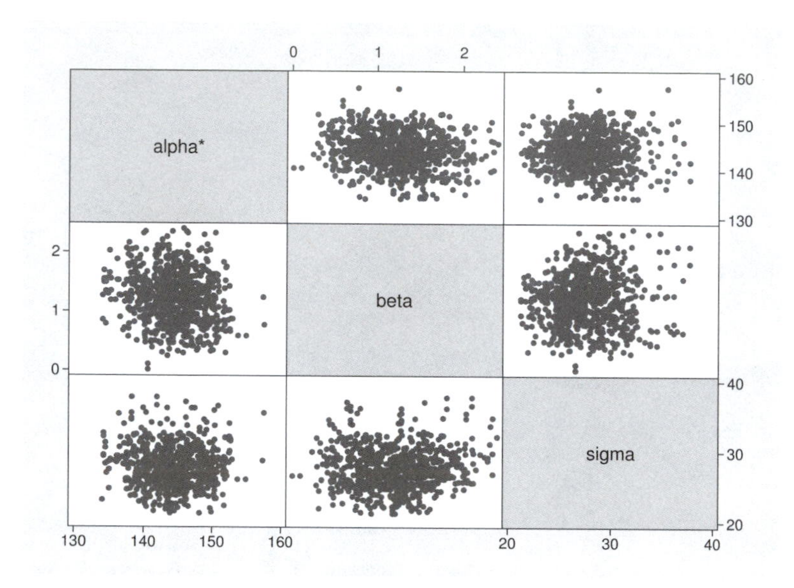

Figure 3.7. Stata scatterplot matrix of the centered analysis of the regression problem

Standard errors of the parameter estimates based on runs of 5,000 simulations are shown in table 3.3. Sigma is unaffected by centering because it is only weakly correlated with the other two parameters. However, the standard errors of both alpha and beta are reduced by more than half when the model is centered, which indicates that the uncentered chain would need to run for more than four times as long to get the same accuracy as a centered chain. Consequently, centering can have a large impact on the efficiency of an MCMC algorithm and should be used routinely whenever a model involves regression on covariates.

Table 3.3. Standard errors of the means (sems) of the regression parameters in centered and uncentered analyses of equal run length

Parameter	Uncentered	Centered
alpha	3.86	1.42
beta	0.07	0.03
sigma	0.09	0.09

3.9.2 Block updating

The short preliminary MCMC run of the uncentered regression model shown in figure 3.6 is sufficient to tell us that the posterior distributions of alpha and beta have standard deviations of about 20 and 0.4, respectively, and that their correlation is about −0.95. This implies an approximate variance matrix for the pair of

$$V = \begin{pmatrix} 400 & -7.6 \\ -7.6 & 0.16 \end{pmatrix}$$

If the model is not to be reparameterized, then it clearly makes sense to update α and β as a pair, and a reasonable proposal distribution for a random walk would be a bivariate normal distribution with this variance matrix. As in the single-parameter case, the exact choice of proposal distribution is not crucial, so even crude guesses at the elements of the variance matrix are likely to improve mixing. A reasonable strategy is to use kV as the variance matrix of the proposal distribution and to adjust k during the burn-in starting with an initial $k = 1$. However, in the regression example, performance with $k = 1$ is more than adequate.

The usual way to generate random values from a multivariate normal distribution is to use the elements of the Cholesky decomposition of the variance matrix as a set of weights for combining independent standard normal random variables. Rather than recalculate the same Cholesky decomposition at every iteration, the **mhsmnorm** command for MH with multivariate normal proposals takes the Cholesky decomposition as one of its inputs in place of the standard deviation of proposal distribution used by the other MH samplers. The code for the regression analysis becomes

begin: block updating

```
matrix b = (80, 0.8, 25)
matrix V = (400,-7.6\-7.6,0.16)
matrix L = cholesky(V)
mcmcrun logpost b using temp.csv, ///
   samp((mhsmnorm, chol(L)) (mhstrnc, sd(0.75) lb(0) ub(50))) ///
   par(alpha beta sigma) burn(1000) update(5000) jpost replace
import delimited temp.csv, clear
```

end: block updating

This algorithm slightly outperformed centering, with these data producing standard errors of 1.25, 0.02, and 0.09 for alpha, beta, and sigma, respectively. Comparison with the results in table 3.3 suggests that the run length for block updating could be about 20% shorter than that for the centered algorithm without loss of accuracy, although some of this difference will be due to not tuning the proposal standard deviations in the two algorithms. Block updating is also slightly faster than cycling through the same parameters: it requires one call to `mhsmnorm` and one log-posterior evaluation as opposed to two calls to `mhsnorm` and two log-posterior evaluations.

3.10 Case study 3: Polyp counts yet again

When fitting the GP model to polyp counts, we saw that the two parameters μ and α are highly correlated and that this slows down the movement of the sampler across the posterior, which means that a very long run was required to adequately represent the distribution. Block updating would be helpful, but both parameters are constrained by their priors to be positive, so a multivariate normal proposal distribution would sometimes propose impossible values and have problems moving around the posterior when it gets close to the parameters' boundaries. One way of handling this problem is to transform the parameters so that they are no longer constrained and then to use a multivariate normal on those transformed parameters. Suppose we transform μ to a new parameter, θ, using $\theta = \log(\mu)$ and that the original prior of μ is $\mathrm{p}(\mu)$; then $\mu = e^{\theta}$, and the Jacobian of the transformation is $|e^{\theta}|$, which makes the transformed prior of θ equal to the following:

$$p(e^{\theta})|e^{\theta}|$$

Thus the log prior under the new parameterization requires the addition of an extra term equal to θ, and the modified program becomes

begin: log transformed block update

```
program logpost
    args logp b

    local mu    = exp(`b´[1,1])
    local alpha = exp(`b´[1,2])
    local theta = `mu´/(1+`alpha´*`mu´)
    tempvar ay lnp
    gen `ay´ = 1+ `alpha´*y
    gen `lnp´ = y*log(`theta´)+(y-1)*log(`ay´)-`theta´*`ay´-log(1-exp(-`theta´))
    quietly summarize `lnp´ [fw=f]
    scalar `logp´ = r(sum)
    logdensity gamma `logp´ `mu´    3 0.3
    logdensity gamma `logp´ `alpha´ 0.5 2
    scalar `logp´ = `logp´ + `b´[1,1] + `b´[1,2]
end

matrix b = (1.0, 0.1)
matrix V = (.11, -0.19 \ -0.19, 0.36)
matrix L = cholesky(V)
mcmcrun logpost b using temp.csv, samp(mhsmnorm, ch(L)) ///
  par(lmu lalpha) burn(1000) update(100000) thin(5) replace jpost
import delimited temp.csv, clear
```

end: log transformed block update

The elements in the matrix V were obtained from the initial short run of the untransformed analysis. The simulated values of mu and alpha were log transformed, and their covariance matrix was found. The precise values in V are not critical and merely have to capture the general shape of the joint posterior. Running this program for 100,000 simulations thinned by 5 produced considerably increased sampling accuracy as reflected in the small standard errors.

```
. mcmcstats mu alpha
----------------------------------------------------------------------------
Parameter             n      mean        sd       sem   median          95% CrI
----------------------------------------------------------------------------
 mu               20000     0.489     0.129    0.0013     0.488 (   0.242,     0.746 )
 alpha            20000     0.837     0.458    0.0050     0.731 (   0.304,     2.015 )
----------------------------------------------------------------------------
```

Comparing these results with the untransformed, brute-force analysis, we see that the sampling variance of alpha has reduced by a factor of $9.2 = (0.0152/0.0050)^2$. Thus a block-updated analysis on the transformed scale based on a run length of 10,000 simulations would provide equivalent precision to the brute-force analysis of 100,000 simulations. Once again, this gain is trivial for this model because it runs quickly but might be critical in a larger model that takes hours to process.

To date, the search for a "black-box" program for Bayesian analysis has met with limited success, although as we will see in chapter 8, WinBUGS often performs well, and improved computing power makes brute-force solutions increasingly practical. While the search for the perfect black box continues, considerable scope remains for ingenuity in designing efficient samplers.

3.11 Starting points

The article by Chib and Greenberg (1995) is probably a sufficient introduction to the MH algorithm for most people, but Brooks (1998) extends this introduction to discuss some of the variations on MCMC. Many specific applications of MCMC are considered in Gilks, Richardson, and Spiegelhalter (1996). The excellent book by Tanner (1996) describes MCMC very concisely and contrasts it with other methods of inference, and Hitchcock (2003) provides a brief history of the development of these algorithms.

For those who want a more mathematical approach, there is Tierney (1994), and because the mathematics behind MCMC algorithms is quite complex, a good starting point for reading about Markov chains is Nummelin (2002).

Roberts and Rosenthal (2001), among many others, discuss the scaling of MH algorithms, and there is a lot of literature discussing the problem of designing adaptive algorithms that change during the actual run without altering the target distribution.

3.12 Exercises

1. Xie and Aickin (1997) analyzed data from a phase III colorectal cancer-prevention trial regarding the ability of wheat bran fiber supplement to prevent the recurrence of colorectal polyps. Subjects with polyps were recruited, and their polyps were removed; the subjects were randomized to extra fiber or no extra fiber and followed up with to see whether the polyps recurred. The data on the women in the trial were analyzed earlier in this chapter. Table 3.4 shows the corresponding polyp counts for the men recruited into the trial.

Table 3.4. Number of polyps removed from men recruited into the cancer-prevention trial

	Number of polyps removed									
	1	2	3	4	5	6	7	8	9	10
Frequency	434	178	87	40	23	12	7	4	1	3

Using a gamma(3,0.3) prior for the average rate, λ, fit the zero-truncated Poisson distribution to the data on the men using an MH algorithm with a log-normal random-walk proposal distribution.

a. Identify a good value for the proposal standard deviation and a suitable length of burn-in.
b. Run your algorithm five times with a run length of 5,000 after the burn-in, and each time note the posterior mean estimate of λ. Are the results consistent to two decimal places? Find the smallest run length that will give two decimal places of accuracy.
c. Check the sensitivity of the estimate of λ to the choice of the prior by contrasting gamma(3,0.3) with gamma(1.5,0.6) and gamma(6,0.15).
d. Analyze the data on men and on women under a zero-truncated Poisson model to decide whether the average rate of polyps varies by gender.
e. The zero-truncated GP model allows for clustering of polyps within susceptible individuals. Investigate this model by running a joint analysis of the data on men and women in which men and women have different means, μ_m and μ_f, but the same amount of clustering, α. Set your own priors. Investigate the posterior of $\mu_m - \mu_f$.

2. Draper et al. (2007) described data for the number of very preterm (VPT) births in 1997 in two regions, Nord-Pas-de-Calais in France and Trent in the United Kingdom. These showed 820 VPT babies in 54,815 live births and 1,149 in 59,394, respectively. VPT was defined as a birth between 20 and 32 weeks' gestation and excluded terminations. In a subsequent paper, Field et al. (2009) reported a survey of 10 European regions and for each gave the number of VPT births in 2003 as shown in table 3.5.

Table 3.5. Birth data for 10 European regions in the 2003 survey

Country	Region	Total births	Total VPT births
Belgium	Flanders	60,444	743
Denmark	Eastern	34,065	381
France	Île-de-France	84,867	1,202
Germany	Hesse	52,078	724
Italy	Lazio	51,939	569
Netherlands	East & Central	48,235	513
Poland	Wielkopolska/Lubuskie	43,188	566
Portugal	Northern	35,336	369
UK	Northern	56,990	949
UK	Trent	30,329	517
Total		497,482	6,533

In the initial analysis, a common probability of VPT will be assumed to apply across the whole of Europe.

a. Use a binomial model for the total number of VPT births in 2003 (6533/497482). Base your prior for the binomial parameter, p, on the 1997 survey, and approximate it by a suitable beta distribution. Use an MH algorithm to derive the posterior for p.

The second analysis estimates the 10 probabilities of VPT for the individual regions without assuming that they are equal. Consider whether you would use the same prior for each region as you used for the average across the whole of Europe. If necessary, modify your prior for use in this new analysis.

b. Prepare a program that estimates within the same iteration of the MH algorithm all 10 region-specific probabilities. Run an MCMC analysis using this program.
c. Estimate the probability that the Trent region of the United Kingdom has the highest rate of VPT births out of the 10 regions.
d. Estimate the probability that the parameter p for Flanders is higher than that for Trent.
e. Estimate the range of p across the 10 regions. Plot a histogram of the estimated range. Does this distribution demonstrate that the values of p are not the same in every region?

4 Gibbs sampling

4.1 Introduction

Section 3.7 described how it is possible to reduce the problem of simulating from a multiparameter posterior to a series of single-parameter steps by cycling through the parameters and applying the Metropolis–Hastings (MH) algorithm to each one in turn. This method works well provided that correlation between parameters is not too high and the proposal distributions are chosen carefully. When the parameters are divided into two subsets θ_1 and θ_2, an ideal MH algorithm uses the conditional posterior distribution at iteration t as the proposal distribution for the next θ_1:

$$q(\theta_1^{(t)}, \theta_1') = p(\theta_1' | y, \theta_2^{(t)})$$

This allows the proposals to range as widely as possible; what is more, the distributions cancel in the MH acceptance probability, which means that every proposal is accepted.

Cycling through the parameters and simulating directly from the conditional posterior distributions without the need for any rejection is called Gibbs sampling. When the parameters are divided into k subsets, the Gibbs sampling algorithm can be summarized as follows:

1. sample $\theta_1^{(t+1)}$ from $p(\theta_1 | y, \theta_2^{(t)}, \theta_3^{(t)}, \ldots, \theta_k^{(t)})$
2. sample $\theta_2^{(t+1)}$ from $p(\theta_2 | y, \theta_1^{(t+1)}, \theta_3^{(t)}, \ldots, \theta_k^{(t)})$

...

k. sample $\theta_k^{(t+1)}$ from $p(\theta_k | y, \theta_1^{(t+1)}, \theta_2^{(t+1)}, \ldots, \theta_{k-1}^{(t+1)})$

Gibbs sampling produces a chain that mixes faster than an MH algorithm with a less well-tuned proposal distribution. However, to use Gibbs sampling, one must be able to simulate directly from the conditional posterior distributions. For some models, this can be achieved by studying the algebra and recognizing when the conditional distributions have a standard form, such as a normal or gamma distribution, for which there is a Stata function for generating the required random values. This produces a highly efficient algorithm in terms of both speed of calculation and convergence. In practice, difficulties arise when some of the conditional distributions are not recognized or when we want to avoid the algebra. For illustration, we will describe an example in which the conditional posterior distributions all take recognizable forms. After that, we will move on to the more realistic case in which some or all conditional distributions are nonstandard.

4.2 Case study 4: A regression model for pain scores

Alund et al. (2000) reported a study of shoulder surgery that included data on preoperative pain and the patients' ages. The data for 24 patients with seropositive rheumatoid arthritis are shown in table 4.1.

Table 4.1. Preoperative pain scores against age for the shoulder surgery study

Age (years)	Pain score	Age (years)	Pain score
30	9.8	55	9.7
36	7.4	58	4.9
44	9.3	69	5.2 8.8 9.4
45	8.9 9.3	61	7.2 7.2 8.4 8.9
47	7.0 8.5	62	7.8
49	8.1	65	2.2
53	8.7	69	4.4 7.6
54	3.4	77	6.5

In this study, pain was measured on a visual analog scale running between 0 (no pain) and 10 (extreme pain), so to remove the effects of the bounded range, it is convenient to transform the pain score before the analysis by using

$$y = \log\left(\frac{\text{pain}}{10 - \text{pain}}\right)$$

The transformed scores are plotted against age in figure 4.1 and show a general decline in reported pain with age.

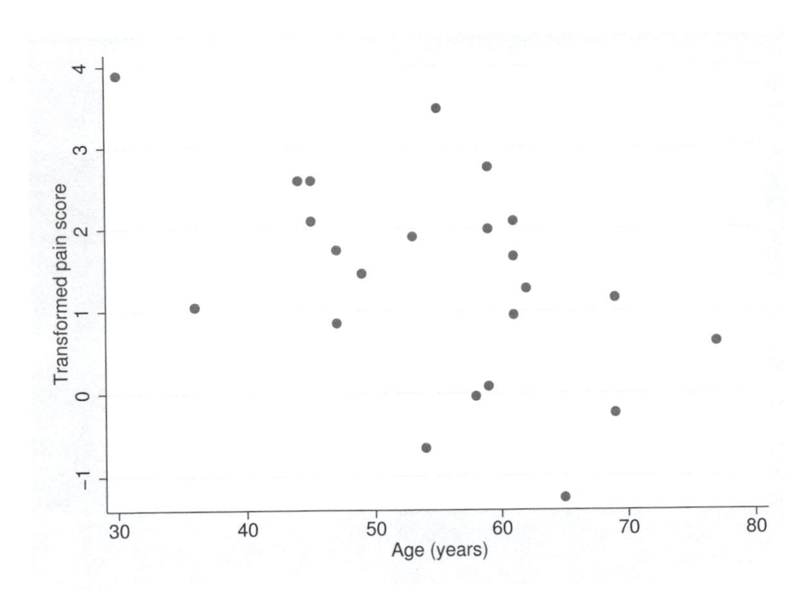

Figure 4.1. Transformed pain score against age

These data will be modeled by a regression in which the average transformed pain score varies linearly with age and the variation about the line is normally distributed with a constant variance. It is common in Bayesian analyses to parameterize the normal distribution in terms of the precision, τ, which is equal to 1 over the variance, because this simplifies the algebra slightly. To improve mixing, we use centering as described in section 3.9.1, so here x denotes `age-55`, and the likelihood can be written as follows:

$$p(y|\alpha, \beta, \tau) = \prod_{i=1}^{24} \sqrt{\frac{\tau}{2\pi}} \exp\left\{-\frac{\tau}{2}(y_i - \alpha - \beta x_i)^2\right\}$$

As is often the case, it is difficult to give informative priors in illustrative problems without being influenced by the data, but we might argue that most of the transformed scores will be in the range -3 to $+3$ and the ages will range from about 30 to 80. Thus an extreme decline will be $-6/50$, and an extreme increase will be $+6/50$. If we have no prior reason to prefer an increase to a decrease, we might opt for a normal prior distribution for β with a mean of 0 and a standard deviation of 0.1 (precision 100). To select a prior for the intercept, α, we must remember that α has been centered, so we are expecting α to be close to the average transformed pain score. To reflect uncertainty in this value, we use an independent noninformative prior that is normal with a mean of 0 and a standard deviation of 10 (precision 0.01). A gamma prior is often used for the precision, τ, and we will select a mean of 1 and standard deviation of 0.5, which implies $G(4, 0.25)$. In summary, the priors are independent and have these forms:

$$p(\beta) = N(0, 0.1) \qquad = \frac{10}{\sqrt{2\pi}} \exp\left(-\frac{100}{2}\beta^2\right)$$

$$p(\alpha) = N(0, 10) \qquad = \frac{0.1}{\sqrt{2\pi}} \exp\left(-\frac{0.01}{2}\alpha^2\right)$$

$$p(\tau) = G(4, 0.25) \qquad = \frac{4^4 \tau^3 \exp(-4\tau)}{\Gamma(4)}$$

Combining these priors with the likelihood and dropping unnecessary constants give an expression for the joint posterior:

$$p(\alpha, \beta, \tau|y) = \exp\left(-50\beta^2\right) \exp\left(-0.005\alpha^2\right) \tau^3 \exp(-4\tau)\tau^{12}$$
$$\exp\left\{-\frac{\tau}{2}\sum_{i=1}^{24}(y_i - \alpha - \beta x_i)^2\right\}$$

Gibbs sampling requires the distribution of τ conditional on knowing the values of α and β. Its equation can be found by selecting the terms that depend on τ in $p(\alpha, \beta, \tau|y)$. In this case, it takes a recognizable form.

$$p(\tau|\alpha, \beta, y) \propto \tau^{15} \exp\left[-\frac{\tau}{2}\left\{8 + \sum_{i=1}^{24}(y_i - \alpha - \beta x_i)^2\right\}\right]$$
$$= G\left[16, 2\left\{8 + \sum_{i=1}^{24}(y_i - \alpha - \beta x_i)^2\right\}^{-1}\right]$$

So to generate a new value of τ, we use the current values of α and β to calculate the two parameters of the gamma conditional distribution, and Stata's `rgamma()` function generates the simulation. Because the conditional distribution has been used to propose the new value, the acceptance probability must be 1, so the proposal can be accepted automatically.

Similar algebra shows that the Gibbs samplers for α and β require normal random variables with means and precisions:

$$p(\alpha|\beta, \tau, y) = N\left\{\frac{\tau\sum_{i=1}^{24}(y_i - \beta x_i)}{0.01 + 24\tau}, 0.01 + 24\tau\right\}$$

$$p(\beta|\alpha, \tau, y) = N\left\{\frac{\tau\sum_{i=1}^{24} x_i(y_i - \alpha)}{100 + \tau\sum_{i=1}^{24} x_i^2}, 100 + \tau\sum_{i=1}^{24} x_i^2\right\}$$

These calculations can be programmed in Stata by placing them in a loop and posting the simulations to a results file. To avoid repeatedly recalculating the same constants, we must calculate the sums and sums of squares of x and y and store them in scalars before running the code. Thus `Sx` contains the sum of the x's and `Sxx` contains the sum of x squared, etc.

begin: Gibbs sampler

```
tempname pf
postfile `pf´ alpha beta tau using temp.dta,replace
scalar alpha = 0
scalar beta = 0
scalar tau = 1
forvalues i=1/2500 {
  post `pf´ (alpha) (beta) (tau)
  scalar b = 0.5*(8+Syy+24*alpha*alpha+beta*beta*Sxx- ///
     2*alpha*Sy-2*beta*Sxy+2*alpha*beta*Sx)
  scalar tau = rgamma(16,1/b)
  scalar t = 0.01+24*tau
  scalar m = (tau*(Sy-beta*Sx))/t
  scalar alpha = rnormal(m,1/sqrt(t))
  scalar t = 100+tau*Sxx
  scalar m = (tau*(Sxy-alpha*Sx))/t
  scalar beta = rnormal(m,1/sqrt(t))
}
postclose `pf´
use temp.dta, clear
generate sd = 1/sqrt(tau)
```

end: Gibbs sampler

Figure 4.2 shows the traces of the parameters, and the mixing appears to be good. Averaging the 2,000 simulations that remain after discarding the first 500 gives a posterior mean α of 4.50 and a posterior mean β of -0.06, so the fitted line is $y = 4.50 - 0.06 \times \text{age}$. As one would expect with such vague priors, the fit of this Bayesian analysis is similar to that given by Stata's `regress` command, `regress y x`.

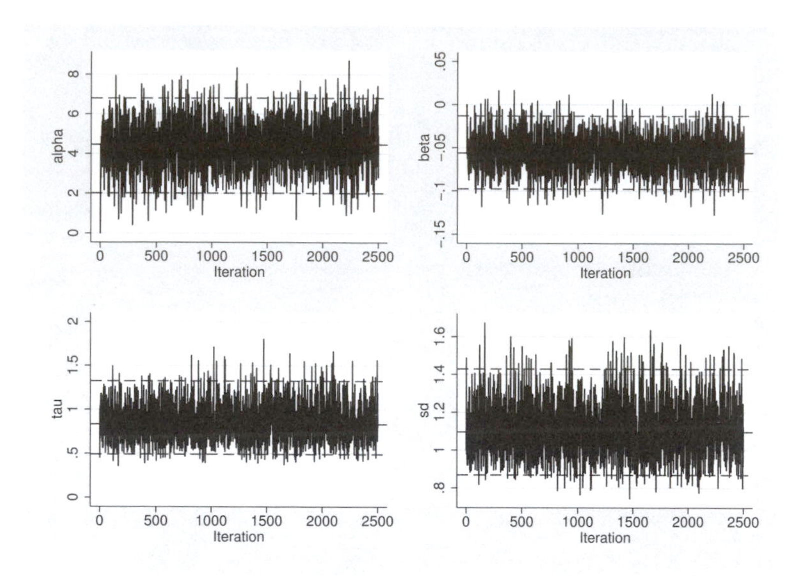

Figure 4.2. Trace plot of the parameters in the regression analysis of the shoulder pain scores

An alternative way of programming this Gibbs sampler is to use the `mcmcrun` command as we did for the `mhs` commands in chapter 3. The difference here is that we never evaluate the log posterior, so we need to replace the reference to the program `logpost` with a nonexistent dummy. The following code implements this. Again we assume that the sums of squares have been saved in scalars.

begin: Gibbs sampling with `mcmcrun`

```
program myGibb
   args dummy b ipar

   local alpha = `b'[1,1]
   local beta  = `b'[1,2]
   local tau   = `b'[1,3]
   if `ipar' == 1 {
      local t = 0.01+24*`tau'
      local m = (`tau'*(Sy-`beta'*Sx))/`t'
      matrix `b'[1,1] = `m' + rnormal()/sqrt(`t')
   }
   else if `ipar' == 2 {
      local t = 100+`tau'*Sxx
      local m = (`tau'*(Sxy-`alpha'*Sx))/`t'
      matrix `b'[1,2] = `m' + rnormal()/sqrt(`t')
   }
   else {
      local a1 = 16
      local a2 = 0.5*(8+Syy + 24*`alpha'*`alpha' + `beta'*`beta'*Sxx ///
        - 2*`alpha'*Sy - 2*`beta'*Sxy + 2*`alpha'*`beta'*Sx)
      matrix `b'[1,3] = rgamma(`a1',1/`a2')
   }
```

```
end
matrix b = (0, 0, 1)
mcmcrun dummy b using temp.csv, samp(3(myGibb)) ///
  burnin(500) updates(2000) par(alpha beta tau) replace
import delimited temp.csv, clear
generate sd = 1/sqrt(tau)
replace alpha = alpha - 55*beta
```

end: Gibbs sampling with mcmcrun

4.3 Conjugate priors

The combination of normally distributed data and either a normally distributed prior for a component of the predictor of the mean or a gamma prior for the precision occur so frequently that it is worth having the general Gibbs sampling formulas for reference. Suppose that there are n observations, y_i, and a model in which

$$\begin{aligned} y_i &\sim N(\mu_i + \beta x_i, \tau) \quad i = 1, \ldots, n \\ \beta &\sim N(m, t) \end{aligned}$$

Here τ and t are the precisions of the normal distributions, μ_i represents the part of the regression equation that does not depend on β, and x_i represents anything that multiplies β. Under these circumstances, the conditional posterior of β will also be normally distributed, and it will have parameters

$$\beta \sim N\left\{\frac{\tau \sum x_i(y_i - \mu_i) + tm}{\tau \sum x_i^2 + t}, \tau \sum x_i^2 + t\right\}$$

A gamma prior for the precision is also conjugate, so if

$$\begin{aligned} y_i &\sim N(\mu_i, \tau) \quad i = 1, \ldots, n \\ \tau &\sim G(a, b) \end{aligned}$$

the conditional posterior of τ will be

$$\tau \sim G\left[a + \frac{n}{2}, \left\{\frac{1}{b} + \frac{1}{2}\sum (y_i - \mu_i)^2\right\}^{-1}\right]$$

where here we use the Stata parameterization of the gamma distribution in which the mean of $G(a, b)$ is ab.

4.4 Gibbs sampling with nonstandard distributions

The algorithm that was derived for the pain score regression is very quick, but it required a lot of preliminary algebra that was only tractable because we chose conjugate priors that led to recognizable conditional posterior distributions. It might be reasonable to live with such restricted priors in a large complex analysis where speed is a critical factor and there is no alternative analysis, but otherwise having the priors dictated in this way is rather limiting.

For most models, there are no priors that lead to a full set of recognizable conditional posterior distributions, so in practice, it will usually be the case that only a few of the conditional posterior distributions will be recognized. In that situation, Gibbs sampling could be used for the recognized forms, and perhaps, the MH algorithm could be used for the other parameters. Gibbs sampling would only be possible for the nonstandard distributions if there were a method of simulating values from a general univariate distribution. The search for such a general method of simulation has received a lot of attention in the literature, and several algorithms have been proposed, of which three will be considered in detail: griddy sampling, slice sampling, and adaptive rejection sampling. Each is implemented within a set of programs with names beginning `gbs`.

4.4.1 Griddy sampling

Because it is relatively easy to draw a random point from under a polygon, one way to sample from a general univariate distribution is to calculate its density at a grid of points and then draw a random point from the approximation to the density formed by constructing a histogram through the points or by joining the points by straight lines. This process is illustrated in figure 4.3 and was named the griddy sampler by Ritter and Tanner (1992), who were the first to formally propose it.

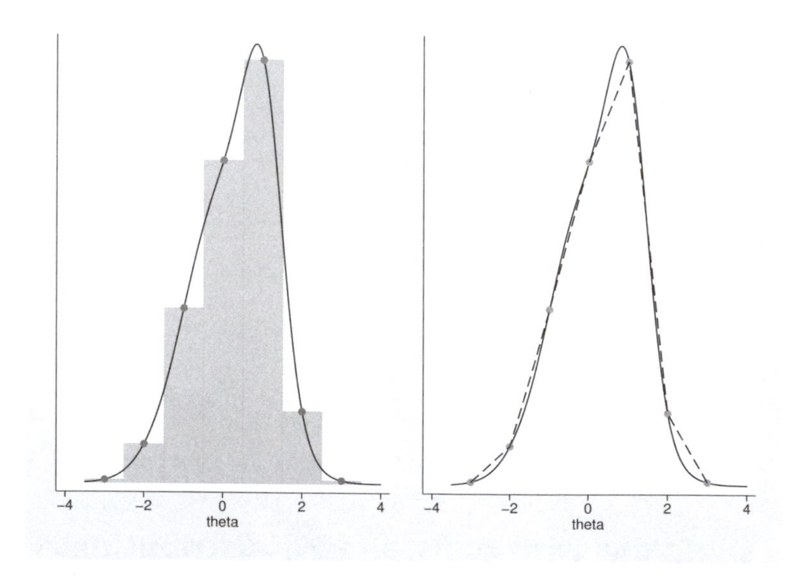

Figure 4.3. Density approximations based on a grid of points

This method has a lot of drawbacks that limit its use in complex problems. Most importantly, it is only an approximation, and the grid might require many points to make it sufficiently accurate; a large grid means many function evaluations and a slow algorithm. It can also be difficult to decide where to place the grid to ensure that the bulk of the distribution is not missed entirely. Nonetheless, the method is easy to program, and it can be made more efficient by adapting the algorithm, perhaps by increasing the number of points once the Gibbs sampler approaches convergence, or by adjusting the spacing of the points in the grid so that they are concentrated in areas of high density.

So long as the grid-based polygon is placed without reference to the current position of the chain, one can think of it not as an approximation to the conditional posterior but as a well-chosen proposal distribution for an MH algorithm. Having generated a point using the griddy sampler, we can subject the value to MH acceptance testing rather than automatically accepting it. Adding this MH step creates samples from the correct distribution rather than from an approximation; the resulting algorithm can have an efficiency close to that of Gibbs sampling provided that the grid-based proposal distribution is close to the corresponding conditional distribution so that there are few rejections.

4.4.2 Slice sampling

Neal (2003) introduced this intuitively simple way of generating values from a general distribution. Given any density $y = f(\theta)$, such as that shown in the left side of figure 4.4, the method seeks to sample a random point (θ, y) within the shaded area, after which y can be discarded and θ will be a random point drawn from $f(\theta)$. In this context, y is an auxiliary variable used to help generate θ, but it is not itself of interest.

The process of sampling from the area under the distribution is illustrated in the right-hand plot of figure 4.4. First, a starting value, θ_0, as shown by the vertical dashed line, is used to draw a random height, y_1, by uniformly sampling in the range, $(0, f[\theta_0])$. Given y_1, the next point, θ_1, is obtained by random sampling within the slice through the shaded area at y_1. Generally, this second stage is time consuming because of the difficulty of finding the ends of the slice. The simplest solution is to ensure detailed balance between the chance of moving from θ_0 to θ_1 and from θ_1 to θ_0 because this will create a Markov chain. Successive points will not be independent, but the chain will converge to the required distribution.

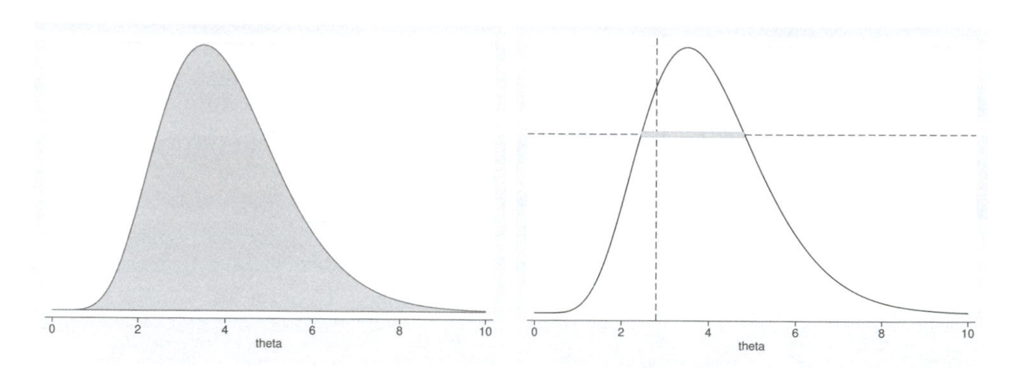

Figure 4.4. Using a slice to select a random point under a distribution

An efficient approach to the second stage of sampling from within the slice is to make an initial guess at the limits of the slice, $[L, R]$, that is wider than necessary and to draw a random θ' from within this range. If $f(\theta') > y_1$, the point lies within the slice and $\theta_1 = \theta'$; if not, the interval $[L, R]$ can be shrunk. When the rejected point is above θ_0, then the upper limit, R, can be reduced to θ' before resampling; when the rejected point is less than θ_0, it replaces L. By shrinking the interval, one reduces the chance of future points lying outside the slice.

Neal (2003) describes one form of this algorithm that requires the following as input: the previous value, θ_0; a guess at a suitable slice width, w; and the maximum number of extensions, m, to the slice that will be tried. The algorithm then calculates limits $[L, R]$ enclosing the slice by stepping out from the initial guess, and subsequently rejected points are used to narrow the interval enclosing the slice.

begin: Slice sampling algorithm

1. $U_1 \sim \text{Uniform}(0, 1),\ y_1 = U_1 f(\theta_0)$
2. $U_2 \sim \text{Uniform}(0, 1),\ L = \theta_0 - wU_2,\ R = L + w$
3. $U_3 \sim \text{Uniform}(0, 1),\ J = \text{int}(mU_3),\ K = (m-1) - J$
4. While $J > 0$ & $y < f(L)$ $L = L - w,\ J = J - 1$
5. While $K > 0$ & $y < f(R)$ $R = R + w,\ K = K - 1$
6. $U_4 \sim \text{Uniform}(0, 1),\ \theta' = L + U_4(R - L)$
7. If $y_1 < f(\theta')$, put $\theta_1 = \theta'$ and stop
8. Otherwise, if $\theta' < \theta_0$, replace $L = \theta'$ and return to step 6
9. Otherwise, replace $R = \theta'$ and return to step 6

end: Slice sampling algorithm

Neal (2003) also describes a number of variations on this algorithm and discusses its generalization to higher dimensions. The basic algorithm as given above is available in the program `gbsslice`.

4.4.3 Adaptive rejection

Rejection sampling is a simple method for generating a random value from a given distribution. Suppose $f(\theta)$ is a function that is proportional to the required density and we can find a function $g(\theta)$ that always exceeds $f(\theta)$ but which is itself easy to sample from. This situation is pictured in figure 4.5. The rejection method involves sampling a random value from $g(\theta)$ and accepting it with probability $f(\theta)/g(\theta)$. If a proposal is rejected, the process is repeated until an acceptable point is found.

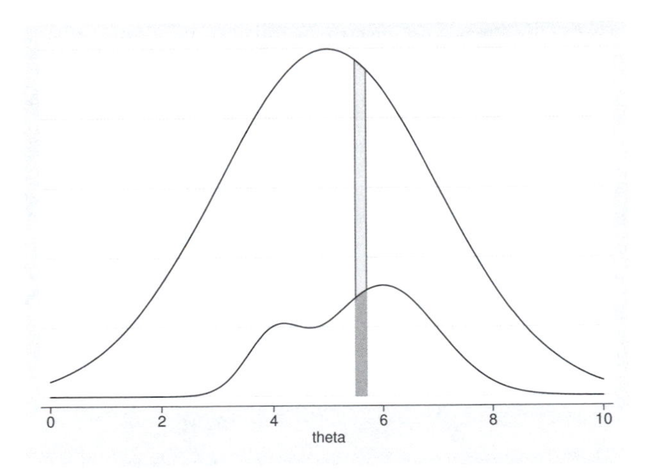

Figure 4.5. Rejection sampling from the lower function, $f(\theta)$, using values randomly generated from the upper function, $g(\theta)$

Rejection sampling is attractively simple, but it is has two drawbacks that limit its use. First, only functions that are easy to sample from can be used as the enveloping function, $g(\theta)$, and second, we must guarantee that $g(\theta)$ is larger than $f(\theta)$ for all possible parameter values. Even worse, $g(\theta)$ must not be too much larger than $f(\theta)$; otherwise, their ratio will be small, and suggested values will rarely be accepted, which will make the algorithm inefficient. These conditions mean that it is difficult to find a suitable function $g(\theta)$ without studying the algebra of $f(\theta)$, and methods that require algebraic tuning are not suitable for programming into a general routine.

Gilks and Wild (1992) suggested a neat way of constructing an enveloping function $g(\theta)$ based on an initial set of trial points. There are several variations on the basic method, but they only guarantee that $g(\theta)$ is always greater than $f(\theta)$ when we are sure that $f(\theta)$ is log concave. The requirement of log concavity is also a major limitation because it requires us to check the algebra of $f(\theta)$ before using the algorithm. Nonetheless, the method is interesting, and as we will see, it does generalize to a form that can be applied to any function.

Figure 4.6 shows how an enveloping function can be created for log-concave functions. The diagram shows $\log\{f(\theta)\}$ and four trial points. Pairs of consecutive points are jointed by straight lines, and these lines are used to create a polygonal envelope, shown in the diagram by the bolder portions of the lines. Because the lines are straight on the log scale, they represent piecewise exponential functions enveloping $f(\theta)$.

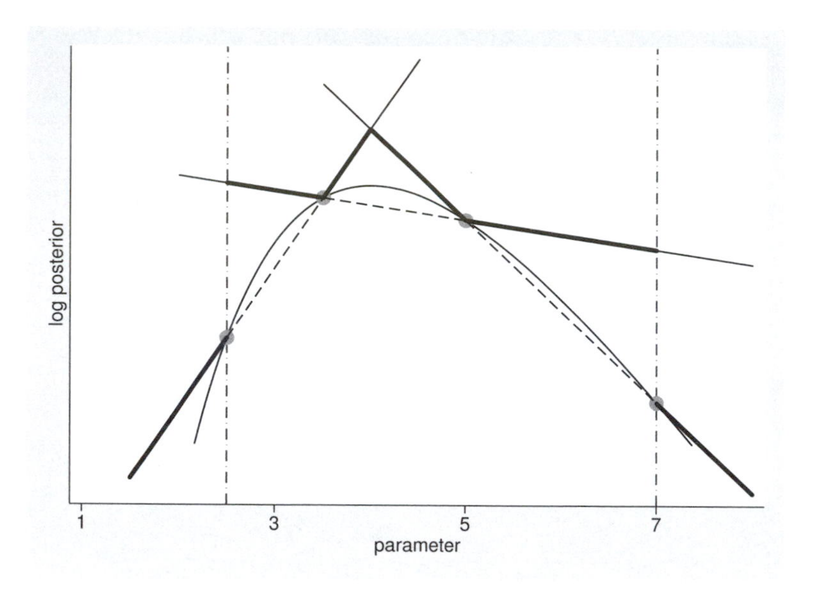

Figure 4.6. An envelope created from four fixed points for use in a rejection sampling algorithm

It is easy to sample from a piecewise exponential, so the algorithm proceeds as follows:

begin: rejection sampling algorithm

1. Select a set of trial points, and construct the polygonal envelope
2. Select a random point θ' from the piecewise exponential, $g(\theta)$, implied by the envelope
3. Evaluate $g(\theta')$ and $f(\theta')$, and accept θ' if

$$U < \frac{f(\theta')}{g(\theta')} \quad \text{where} \quad U \sim U(0,1)$$

4. Otherwise, return to step 2

end: rejection sampling algorithm

In fact, this algorithm can be made slightly more efficient by first checking the value of the dashed interior line at our proposed parameter value. Because of the log concavity, this value, $\log\{h(\theta')\}$, will be less than $\log\{f(\theta')\}$, and the proposed value can be accepted whenever

$$U < \frac{h(\theta')}{g(\theta')}$$

In this way, it will sometimes be possible to accept a proposal without the more time-consuming calculation of $f(\theta')$.

The efficiency of the Gilks and Wild (1992) rejection algorithm depends on the choice of the initial points used to construct the envelope. Gilks and Wild (1992) suggest making the algorithm adaptive by adding any rejected point, θ', to the fixed set of trial points and recalculating the envelope. In this way, a poor choice of initial points may lead us to reject the proposal, but the rejected point will be added to the grid thus improving subsequent proposals.

Clever as the adaptive algorithm is, it remains a rejection method and only works if it can be guaranteed that the envelope is always larger than the function being sampled. That is why the function must be concave on a log scale. However, as we did with the griddy sampler, any rejection method can be adapted so that $g(\theta)$ is not used as an envelope but as a proposal distribution for an MH sampler. If $g(\theta)$ is close to $f(\theta)$, this should lead to a very efficient algorithm. Indeed, as $g(\theta)$ approaches $f(\theta)$, the method approaches Gibbs sampling, in which all proposals are automatically accepted. Gilks, Best, and Tan (1995) and Gilks et al. (1997) developed this idea to create an adaptive MH sampler.

The MH version of adaptive rejection sampling (ARS) is known as adaptive rejection Metropolis sampling (ARMS), and starting from the current value, $\theta^{(t)}$, it involves as its first stage a straightforward rejection sampling algorithm. However, because for a general density, we cannot assume that $g(\theta) > f(\theta)$, rejection sampling will sample points with probability $f(\theta)$ whenever $g(\theta) > f(\theta)$, and it will sample points with probability $g(\theta)$ whenever $g(\theta) < f(\theta)$. In effect, the point, θ', that is eventually accepted comes from a function constructed as $\min\{g(\theta), f(\theta)\}$. If the sampled point is subjected to an MH step, then to converge to $f(\theta)$, it needs an acceptance probability equal to

$$\alpha(\theta^{(t)}, \theta') = \min\left[\frac{f(\theta')\min\{f(\theta^{(t)}), g(\theta^{(t)})\}}{f(\theta^{(t)})\min\{f(\theta'), g(\theta')\}}, 1\right]$$

In summary, the ARMS algorithm becomes

begin: ARMS algorithm

1. Create the piecewise exponential envelope, $g(\theta)$
2. Randomly sample θ' from $g(\theta)$
3. Generate $U \sim \text{Uniform}(0,1)$
4. If $U > f(\theta')/g(\theta')$, reject θ', add it to the grid, and return to step 1
5. Otherwise, run the MH step with acceptance probability

$$\min\left[1, \frac{f(\theta')\min\{f(\theta^{(t)}), g(\theta^{(t)})\}}{f(\theta^{(t)})\min\{f(\theta'), g(\theta')\}}\right]$$

end: ARMS algorithm

This algorithm is not unlike a Metropolized version of the griddy sampler except that the initial rejection step means that the proposed points will always be accepted in the MH step whenever both the previous and new points come from regions where $g(\theta) > f(\theta)$.

Gilks, Best, and Tan (1995) suggest an alternative piecewise exponential function that lies closer to a more general function without always lying above it,

$$\max\left\{L_{i,i+1}, \min\left(L_{i-1,i}, L_{i+1,i+2}\right)\right\}$$

where $L_{i,j}$ is the line through points i and j of the grid.

ARS and ARMS both work well provided that the initial grid is close to the region of high-posterior density. However, if the grid is poorly placed, the algorithm may fail to identify a region of high density outside the original range, in which case, it may continue to select points near the center of the misplaced grid. In this way, the algorithm becomes stuck in regions that have poor posterior support. The usefulness of the grid-based algorithms depends critically on whether you can set an appropriate grid. To be safe, one needs to use an initial grid with many points to make sure that the well-supported regions are covered; however, this defeats the object because the grid would require many evaluations of the log posterior and hence become inefficient. Indeed, unless you can place a small grid with some certainty, grid-based algorithms tend to perform badly.

4.5 The gbs commands

`gbsgriddy`, `gbsslice`, `gbsars`, and `gbsarms` form a set of samplers with similar syntax that are suitable for use in a Gibbs sampler. The command name is followed by the name of the program that evaluates the log posterior, the row vector containing the parameters, and the number of the parameter to be updated. The allowed options vary with the command and are given in the help files. The general syntax takes the following form:

`gbs`*cmd logpost rowvector parameter* [*if*] [*in*] [*weight*] [, *options*]

Most algorithms are based on an initial grid of points placed by the user to cover the region where the posterior distribution is thought to concentrate. This grid must be specified by the user. If a sparse grid is made too wide, the region of high-posterior probability could fall between two successive points and be missed. Several algorithms include an MH step, and the grids for those algorithms must not be placed on the basis of previous values from the same chain, because of the theoretical possibility that such an adjustment would alter the distribution to which the chain converges. As with the adjustment of the proposal standard deviation for the MH routines, it would be legitimate to modify the grid during the discarded burn-in provided that the grid is not further adapted on the basis of the actual run.

4.6 Case study 4 continued: Laplace regression

Consider again the analysis of the association between shoulder pain and age that is illustrated in figure 4.1, but instead of normal errors regression, let us use a distribution with much longer tails to make the regression more robust to outliers. One possibility is the Laplace distribution, sometimes called the double exponential distribution. Under this model, the likelihood is

$$p(y|\alpha,\beta,\tau) = \prod_{i=1}^{24} \frac{1}{2\phi} \exp\left(-\frac{|y_i - \alpha - \beta x_i|}{\phi}\right)$$

Assuming that our prior beliefs about α and β have not changed, all that is needed to complete the model is a prior for ϕ. The standard deviation of the Laplace distribution is $\sqrt{2}\phi$, so rather than trying to place a prior on ϕ, we will use a prior for the standard deviation that is uniform over the range (0,5). Finally, let us suppose that we are too lazy to attempt the algebra needed to see whether any of the conditional distributions take a recognizable form and instead accept the slightly longer run times required by griddy sampling, slice sampling, and adaptive rejection.

It is simple to write a program that evaluates the product of likelihood and prior. The program below is written for simplicity rather than speed but does incorporate centering because this greatly improves the mixing.

begin: logpost

```
program logpost
    args logp b

    local alpha = `b´[1,1]
    local beta  = `b´[1,2]
    local sigma = `b´[1,3]
    local phi   = `sigma´/1.414214
    scalar `logp´ = 0
    logdensity laplace `logp´ y `alpha´+`beta´*x `phi´
    logdensity normal   `logp´ `alpha´ 0 10
    logdensity normal   `logp´ `beta´  0 0.1
    logdensity uniform  `logp´ `sigma´ 0 5
end
```

end: logpost

A much more efficient version of the same program can be obtained by explicitly programming the algebraic form of the log posterior.

begin: efficient logpost

```
program logpost
    args logp b

    local alpha = `b´[1,1]
    local beta  = `b´[1,2]
    local sigma = `b´[1,3]
    local phi = `sigma´/1.414214
    tempvar u
    generate `u´ = -abs(y-`alpha´-`beta´*x)/`phi´ - log(`phi´)
    quietly summarize `u´, meanonly
    scalar `logp´ = r(sum) - 0.005*`alpha´*`alpha´ - 50*`beta´*`beta´
end
```

end: efficient logpost

Ten thousand calls to the less efficient version of `logpost` took 7.5 seconds, while on the same computer, 10,000 calls to the more efficient version took 0.78 seconds. The efficient code is faster by a factor of about 10 and so is used in all subsequent comparisons. Using a grid of 9 points for each of the three parameters and creating a chain of length 5,000 involves 135,000 evaluations of the log posterior; thus ignoring overheads associated with calculating polygons, sampling points, and writing to the file, we see that the calculations for a chain of length 5,000 take between 10 and 15 seconds. To compare the different algorithms, we use code similar to that given below for the griddy sampler.

begin: griddy sampling

```
matrix b = (1,0,1)
mcmcrun logpost b using temp.csv,                        ///
  samp((gbsgriddy, grid(-1 1 2 3 5))                     ///
       (gbsgriddy, grid(-0.2 -0.1 -0.05 0 0.1))          ///
       (gbsgriddy, grid(0.001 0.5 1 2 5)))               ///
  burn(500) update(2000) par(alpha beta sigma) replace
import delimited temp.csv, clear
replace alpha = alpha - 55*beta
generate phi = sigma/1.414214
```

end: griddy sampling

Notice that the lower bound for the standard deviation is placed at 0.001 even though the theoretical range is (0,5). If we were to use zero as the lower bound, then the grid-based algorithms would attempt to evaluate the log posterior at zero and fail because the calculation would involve division by zero. The exact placement of the lower limit might have to be revisited if we were to find that the posterior lies close to zero, but in this case, that was not a problem. In this example, the initial values and limits for the first sampler apply to the uncentered parameter α.

To create a benchmark, we first ran the Laplace regression analysis using the MH algorithm described in chapter 3 with adaptation of the proposal standard deviation during the burn-in. Following that, we reran the program with different samplers, and the results are shown below.

```
Metropolis-Hastings ( 0.74s per 1,000 iterations)
----------------------------------------------------------------------------
Parameter          n       mean         sd        sem    median          95% CrI
----------------------------------------------------------------------------
 alpha          5000      4.465      1.261     0.0398     4.501 (   1.855,     6.863 )
 beta           5000     -0.053      0.022     0.0007    -0.054 (  -0.098,    -0.009 )
 phi            5000      0.961      0.205     0.0064     0.930 (   0.638,     1.442 )
----------------------------------------------------------------------------
Griddy Sampling: 9 point grids ( 3.49s per 1,000 iterations)
----------------------------------------------------------------------------
Parameter          n       mean         sd        sem    median          95% CrI
----------------------------------------------------------------------------
 alpha          5000      4.375      1.578     0.0223     4.358 (   0.977,     7.743 )
 beta           5000     -0.053      0.027     0.0004    -0.051 (  -0.115,     0.005 )
 phi            5000      1.077      0.318     0.0049     1.029 (   0.637,     1.858 )
----------------------------------------------------------------------------
Griddy Sampling: 9 point grids, histograms ( 3.48s per 1,000 iterations)
----------------------------------------------------------------------------
Parameter          n       mean         sd        sem    median          95% CrI
----------------------------------------------------------------------------
 alpha          5000      4.371      1.578     0.0227     4.400 (   0.821,     7.645 )
 beta           5000     -0.053      0.028     0.0004    -0.052 (  -0.114,     0.008 )
 phi            5000      1.076      0.318     0.0049     1.031 (   0.634,     1.849 )
----------------------------------------------------------------------------
```

Griddy Sampling: 5 point grids, Metropolis (2.34s per 1,000 iterations)

Parameter	n	mean	sd	sem	median	95% CrI
alpha	5000	4.421	1.054	0.0212	4.403	(2.362, 6.746)
beta	5000	-0.052	0.018	0.0004	-0.052	(-0.096, -0.018)
phi	5000	0.936	0.194	0.0042	0.912	(0.645, 1.448)

ARMS: 3 point initial grids (2.95s per 1,000 iterations)

Parameter	n	mean	sd	sem	median	95% CrI
alpha	5000	4.482	1.181	0.0174	4.463	(2.180, 6.749)
beta	5000	-0.054	0.021	0.0003	-0.053	(-0.096, -0.014)
phi	5000	0.933	0.184	0.0027	0.915	(0.632, 1.328)

ARMS: 9 point initial grids (3.89s per 1,000 iterations)

Parameter	n	mean	sd	sem	median	95% CrI
alpha	5000	4.476	1.258	0.0182	4.512	(1.921, 6.802)
beta	5000	-0.054	0.022	0.0003	-0.053	(-0.096, -0.010)
phi	5000	0.969	0.220	0.0032	0.936	(0.628, 1.489)

Slice Sampling: steps approx. 1sd (2.24s per 1,000 iterations)

Parameter	n	mean	sd	sem	median	95% CrI
alpha	5000	4.615	1.151	0.0212	4.606	(2.371, 6.857)
beta	5000	-0.056	0.020	0.0004	-0.055	(-0.098, -0.018)
phi	5000	0.931	0.193	0.0038	0.907	(0.625, 1.373)

Slice Sampling: steps approx. 2sd (2.17s per 1,000 iterations)

Parameter	n	mean	sd	sem	median	95% CrI
alpha	5000	4.612	1.183	0.0196	4.601	(2.286, 6.933)
beta	5000	-0.056	0.021	0.0003	-0.055	(-0.098, -0.016)
phi	5000	0.949	0.210	0.0037	0.923	(0.624, 1.438)

Slice Sampling: steps approx. 4sd (2.13s per 1,000 iterations)

Parameter	n	mean	sd	sem	median	95% CrI
alpha	5000	4.549	1.193	0.0184	4.521	(2.274, 6.870)
beta	5000	-0.055	0.021	0.0003	-0.054	(-0.098, -0.017)
phi	5000	0.957	0.213	0.0035	0.929	(0.621, 1.433)

Several points are immediately evident from this comparison. The initial impression is that the MH algorithm requires fewest evaluations of the log posterior and that it is by far the fastest algorithm. However, this is somewhat misleading. The autocorrelation in the resulting chain means that the standard errors of the mean estimates are approximately twice those of the Gibbs sampling algorithms, which implies that the MH algorithm needs to be run for about four times as long to give the same accuracy. This raises the run length to 20,000 and the total run time to about 15 seconds.

The griddy samplers with 9 points only give approximate answers, but the mean estimates are still very close to the answers given by the exact algorithms. Unfortunately, the same cannot be said for the credible intervals (CrI), and if these are of interest, then these approximate methods need to be based on impractically large grids before they will be accurate enough. Reducing griddy sampling to just five points and including an MH step gives a performance similar to that of the MH with a tuned normal proposal both in terms of the mean and the credible intervals.

ARMS is slightly slower than the Metropolized griddy sampler but gives smaller standard errors so that its performance is broadly comparable. Starting ARMS with 3 points is better than starting with 9 points even though the smaller grid will need to add more points before it finds one that it can accept.

Overall, the slice sampler performs well: it is quick and gives low standard errors. The performance of slice sampling is not seriously sensitive to the choice of step length provided that the choice for each parameter reflects the likely range of its posterior.

It would be valid to mix these samplers in any combination that makes the final program efficient. We might, for instance, use standard Gibbs sampling for recognized distributions, ARS for nonstandard distributions that are known to be log concave, and slice sampling for the remainder. However, here are some recommendations:

- Restrict the use of `logdensity` to small problems or to trial runs of larger problems intended to establish an approximate solution. Efficient programming of the log posterior should be the norm.
- Where possible, identify conditional posterior distributions that have standard forms, and use Gibbs sampling for those parameters.
- ARS is worth considering if you know in advance that the log posterior is concave, and you know roughly where to place the initial grid of points.
- For nonstandard posterior distributions, slice sampling is often the most efficient algorithm, but a well-tuned MH algorithm can be almost as good.
- For highly correlated parameters with an unrecognized joint distribution, use the block updated MH algorithm.
- All samplers struggle with bimodal posterior distributions because they can become stuck under one mode. These arise most commonly from overparameterized models or from informative priors that conflict with sparse data. If bimodal distributions are suspected, try starting the chain from different initial values to see whether the results are consistent.

4.7 Starting points

Casella and George (1992) provide a readable introduction to Gibbs sampling. The algorithm is widely used when the algebra is tractable, with countless articles describing applications to specific models; early articles by Dellaportas and Smith (1993) and

Wakefield et al. (1994) illustrate the method with simple examples. General overviews of Gibbs sampling are provided by many textbooks, including those by Gilks, Richardson, and Spiegelhalter 1996, Tanner (1996), Gelman et al. (2004), and Suess and Trumbo (2010). The griddy sampler was introduced by Ritter and Tanner (1992), but it is not widely used. Neal (2003) introduced the slice sampler and described several variations on the basic algorithm. ARS was suggested by Gilks and Wild (1992) and extended to ARMS by Gilks, Best, and Tan (1995). The original article on ARMS suggested that the starting grid can be based on the final grid from the previous stage, which is not true, as they later explained in Gilks et al. (1997).

4.8 Exercises

1. In their article on the griddy sampler, Ritter and Tanner (1992) analyze data taken from Marske (1967) on the biochemical oxygen demand (BOD) of water samples that were taken from a stream, inoculated with a culture of microorganisms, and left for between one and seven days. The data are given in table 4.2.

Table 4.2. Biochemical oxygen demand (BOD) after different times

Time (days)	BOD (mg/L)
1	8.3
2	10.3
3	19.0
4	16.0
5	15.6
7	19.8

Analyze the data using a nonlinear regression,

$$y_i = \theta_1\{1 - \exp(-\theta_2 t_i)\} + \epsilon_i$$

where y is the BOD measurement, t is the time, and the error structure is normal with a constant variance σ^2.

a. Set priors for the parameters of the model without reference to the actual data, but instead use the following information taken from Wikipedia:

> Most pristine rivers will have a 5-day carbonaceous BOD below 1 mg/L. Moderately polluted rivers may have a BOD value in the range of 2 to 8 mg/L. Municipal sewage that is efficiently treated by a three-stage process would have a value of about 20 mg/L or less. Untreated sewage varies, but averages around 600 mg/L in Europe and as low as 200 mg/L in the U.S.

b. Write a program to evaluate the log posterior of your model, and then compare the performance of an MH algorithm with the griddy, slice, and ARMS versions of Gibbs sampling.

2. Consider again the hierarchical model used in section 2.5 to estimate the relationship between blood pressure and age in people with typical angina. Set a_s and b_s to appropriate constants of your own choosing. Place your own higher-level priors on μ_a, σ_a, μ_b, and σ_b so that each group's regression coefficients are linked to those of the other groups.

   a. Fit the model using a Gibbs sampler with slice sampling for each of the parameters.
   b. Investigate the algebra to see whether any posterior distributions take recognizable forms, and build those distributions into your Gibbs sampler.
   c. Compare the marginal estimate of the regression slope, b_1, for the typical angina group with the corresponding nonhierarchical estimate obtained when only the data for the four patients with typical angina are used.
   d. Fit a model assuming that all four groups have the same standard deviation. Is this assumption reasonable? Does it alter the posterior distribution of b_1?

5 Assessing convergence

5.1 Introduction

Before reporting any findings based on a set of Markov chain Monte Carlo (MCMC) simulations, one must investigate the convergence to confirm that the chain adequately covers the posterior. Unfortunately, there is no way to guarantee such convergence because even after millions of iterations, it is always possible that some remote part of the joint posterior distribution has not yet been visited. This is more of a theoretical possibility than a practical problem, and we are usually concerned with two questions:

- Has enough of the early part of the chain been discarded to remove the influence of the initial values?
- If so, has the chain been run for long enough to capture the important characteristics of the posterior with sufficient accuracy?

Assessing the impact of the initial values is essentially a matter of detecting early drift: once the chain reaches the center of the distribution, the starting values will have no further influence. The second issue of whether a chain has been run for long enough is somewhat more difficult to resolve because it depends on what characteristics of the posterior are required and for what they will be used. Generally, it is easy to obtain accurate estimates of the mean of the posterior distribution but much more difficult to capture the properties of the tails because only a small proportion of the simulations lies in those tails.

Most methods for assessing convergence concentrate on the behavior of the marginal distributions of the individual parameters, but one should remember that the posterior distribution is almost always multidimensional. The mean of one parameter might appear quite stable early in the chain when other parameters have not yet converged, only for that marginal distribution to change once the other parameters reach their true levels. In general, no single parameter can be relied upon until all the parameters have converged.

5.2 Detecting early drift

A trace or history plot of the type provided by `mcmctrace` is the first tool for detecting early drift. This graph plots consecutive values of the chain together with horizontal lines denoting the median and 95% limits. The plot is useful for detecting drift and investigating the quality of the mixing. Ideally, the trace plot of the chain will move quickly across its range without any drift in its mean. If drift is suspected, then the affected portion of the chain should be discarded.

To illustrate the use of the trace plot, we generate a sequence of values with known characteristics to mimic MCMC simulations. The following code creates a chain of length 1,000 with a mean, `mu`; a standard deviation, `sigma`; and correlation between successive values of `rho`, but no early drift. All the artificial MCMC chains shown in this chapter were produced with the same seed for the random-number generator, but it is worth trying other seeds to create different artificial chains with the same broad characteristics; in doing so, one can see the way in which these plots vary in practice.

begin: artificial MCMC

```
set seed 376015
set obs 1000
local mu = 5
local sigma = 1
local rho = 0
local f = sqrt(1-`rho'*`rho')*`sigma'
generate theta = rnormal(`mu',`sigma') in 1
replace theta = rnormal(`mu'*(1-`rho') + `rho'*theta[_n-1],`f') in 2/1000
```

end: artificial MCMC

To investigate the impact of early drift, we modify the code by adding a further command. To produce a linear drift of 1 unit over the first 200 simulations requires the addition of the following command:

```
replace theta = theta - 1 + _n/200 in 1/200
```

The left-hand side of figure 5.1 shows the trace plot for a series generated with `mu` = 5, `sigma` = 1, `rho` = 0, and no drift. This represents an ideal situation; because there is no drift, there should be no concern over the impact of the initial values, and the lack of correlation means that we have a fast-mixing chain. The right-hand side of the plot shows the corresponding cusum plot as generated by the `mcmccusum` command. This graph is based on a suggestion of Yu and Mykland (1998) and plots the cumulative sum of the differences between the individual simulations and the overall mean of the chain. The cusum plot must, by definition, start and end at zero. A run of values of the chain that are above the overall mean will move the cusum in a positive direction; a run of low values will move the cusum in a negative direction. Cusum plots of autocorrelated sequences can vary a lot, so to help with interpretation, an option of `mcmccusum` can be used to create one or more reference curves. These are shown as dotted lines and represent the cusum for a randomly generated sequence with the same mean, standard deviation, and autocorrelation as the actual data, but without drift. Although these reference curves will not necessarily move in phase with the real data, their range and pattern serve as a guide when deciding whether any drift is present. In this ideal case, there is no suggestion of drift.

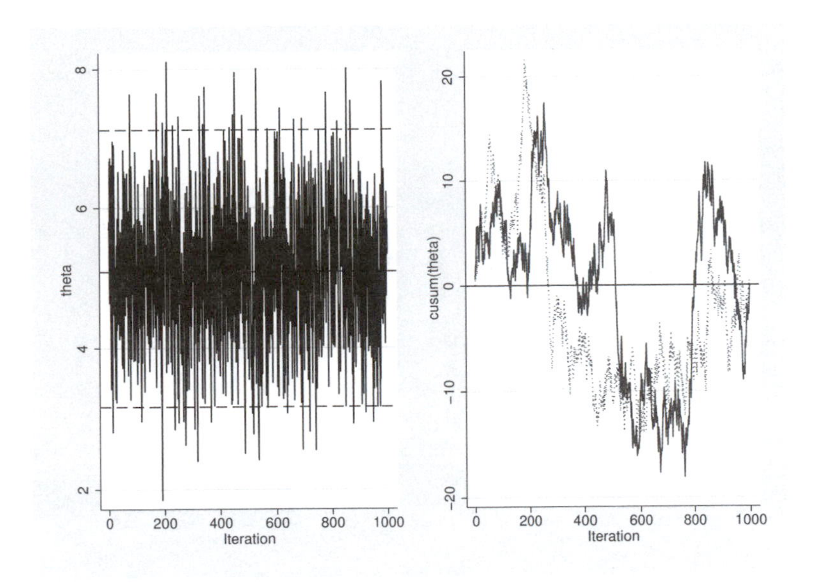

Figure 5.1. Trace and cusum plots for a chain with no autocorrelation and no drift

When a similar chain is generated, again without drift, but with the autocorrelation parameter set to 0.8 to mimic a more typical MCMC algorithm, then figure 5.2 is produced. The slower mixing is evident in both the trace plot, which takes longer to move across its range, and the cusum plot, which has longer runs of positive and negative values and a less spikey appearance. Note also the wider range of the cusum scale in figure 5.2 compared with figure 5.1. Although mixing is worse, there is still no indication of drift.

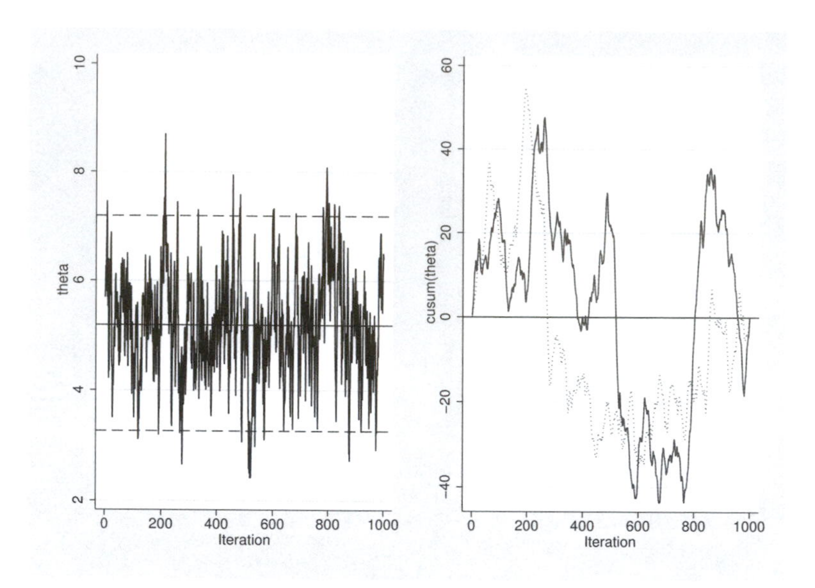

Figure 5.2. Trace and cusum plots for a series with autocorrelation 0.8 and no drift

Adding drift during the first 200 simulations produces figures 5.3 and 5.4 for uncorrelated and autocorrelated (`rho` = 0.8) chains. The drift here is noticeable and easily detected by both plots. Often more subtle drift is easier to see in the cusum plot.

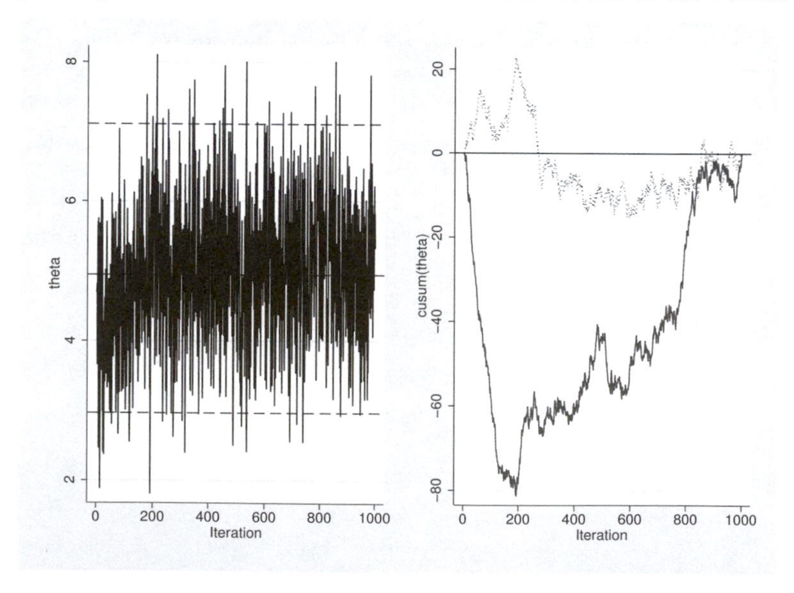

Figure 5.3. Trace and cusum plots for a series with no autocorrelation but drift over the first 200 simulations

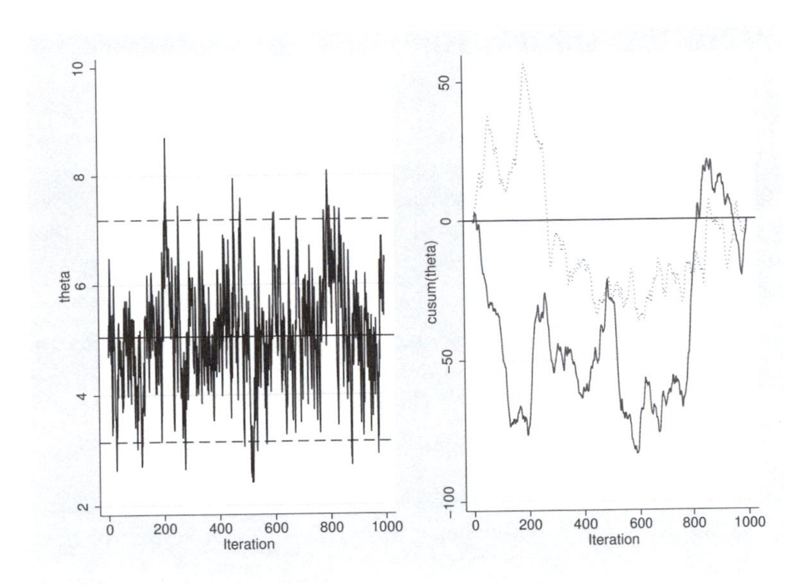

Figure 5.4. Trace and cusum plots for a series with autocorrelation 0.8 and drift over the first 200 simulations

A popular alternative way of detecting early drift is to use a frequentist test that compares the mean of the initial part of the chain with the mean of the final part of the chain. Such a test was suggested by Geweke (1992), and a version of that test is implemented by `mcmcgeweke`. Below is the output from `mcmcgeweke` for analyzing the chain plotted in figure 5.4. By default, the first 10% of the chain is compared with the last 50%, and in this case, the difference in the means is large and significant and would cause us to suspect that the early part of the chain is influenced by the initial values and should be discarded.

```
. mcmcgeweke theta
------------------------------------------------------------
Parameter  Statistic    First Section        Final Section
------------------------------------------------------------
           size         10.0% (n = 100)      50.0% (n=500)
------------------------------------------------------------
theta      mean             4.6926               5.1714
           st error         0.1551               0.1367
           autocorr         0.6312               0.8019
           mean diff                  0.4788
           se diff                    0.2068
           z                          2.3154
           p-value                    0.0206
------------------------------------------------------------
```

5.3 Detecting too short a run

Once the impact of the initial values has been removed by discarding the early part of the chain, there remains the far trickier problem of deciding how long to run the chain to capture the important properties of the posterior distribution. Occasionally, the trace plot will show a sharp jump in level in the middle of a run, which indicates that the posterior is made up of distinct regions of high probability. However, trace plots are generally not much help in determining whether the run was long enough other than in alerting us to a poorly mixing chain.

A more useful graph is produced by `mcmcsection`, which plots the smoothed marginal density of a parameter. By default, `mcmcsection` plots the density estimate based on the full chain and the two estimates based on the first and second halves of the chain. If the distribution has been adequately captured, then the first and second halves should give very similar densities. The statistic D displayed below the plot measures the maximum difference between the densities from the two halves as a percentage of the maximum height of the combined density. Values of D below about 15% represent agreement that will be adequate for most purposes.

To illustrate the use of `mcmcsection`, we generated two chains with an autocorrelation 0.8 but no drift using the code given in the previous section. One chain was of length 500 and the other of length 5,000. The plots produced by `mcmcsection` for the two chains are shown in figure 5.5. Although there is no drift, it is clear that the shorter chain has not settled down, while the longer chain is much more stable.

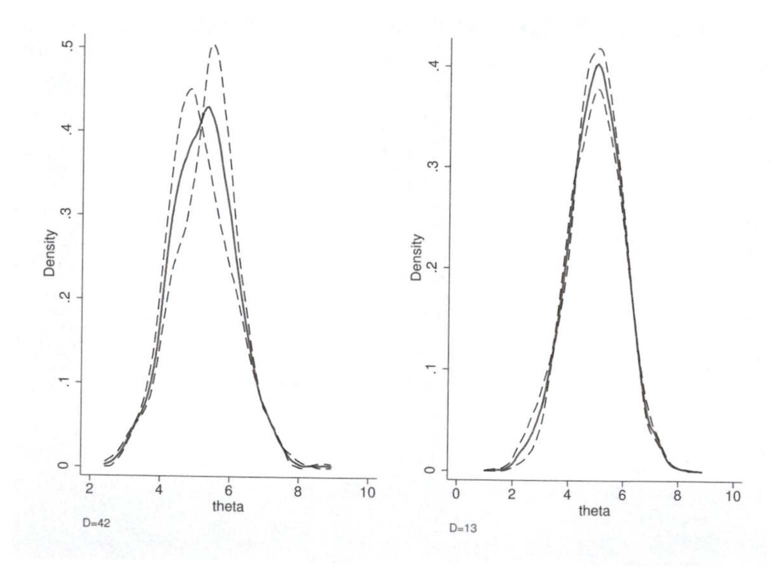

Figure 5.5. Section plots based on 500 and 5,000 simulations from chain with high autocorrelation and no drift

When a single-point estimate of a parameter is needed, Bayesians usually use the mean of the chain, so the adequacy of the length of chain is reflected in the standard error of its mean. The size of the standard error will depend on the pattern of autocorrelation, but it will still be proportional to the square root of the sample size. If a short chain is produced and used to obtain the standard error, the length of run that would be needed to obtain the target accuracy can then be calculated by increasing `n` proportionally. These calculations are performed by the command `mcmclength`. To illustrate them, we created a chain of 5,000 simulations as in figure 5.5. On the basis of this chain, `mcmclength` calculated the following:

```
. mcmclength theta, target(0.025)
theta: target standard error is .025
Current chain length: 5000
Block size for bootstrap: 100
---------------------------------------------------------------------
 range                    estimate     st error       target     updates
---------------------------------------------------------------------
mean                        4.9683       0.0449       0.0250       16130
Standard deviation          1.0001       0.0240       0.0250        4623
95% CrI lower               2.9208       0.0925       0.0250       68450
95% CrI upper               6.8970       0.1024       0.0250       83933
---------------------------------------------------------------------
note: if the current chain is thinned then multiply the projected
 number of updates by that thinning value
```

The default with `mcmclength` is to aim for a standard error that is 5% of the corresponding estimate, although in this example, the default is overridden by specifying the target standard error. The standard errors are estimated by a simple blocked bootstrap procedure, with the default block size chosen by inspecting the autocorrelations and ensuring that the block covers all lagged autocorrelations over 0.05. The results should only be treated as a guide but suggest that to get a standard error due to sampling of 0.025, which is roughly equivalent to one decimal place of accuracy in the estimate, we need a chain of about 16,000 in length for the posterior mean and a chain of about 80,000 in length for the credible interval (CrI). To check these predictions, we created 100 simulated chains of 16,000 in length; the standard deviation of the means of the individual chains was 0.026, very close to the prediction. One hundred chains of length 80,000 gave standard deviations for the limits of the credible intervals of 0.019 and 0.018, suggesting that the predicted run length was rather conservative. Part of the problem here is the approximate nature of this type of bootstrap, but more critical is the sample used to create the predictions. The seed of 376015, chosen arbitrarily for use in the examples presented in this chapter, happens to give an initial sample that suggests that uncharacteristically large run lengths will be needed. By trying other seeds, we see that a more typical predicted run length is about 40,000 iterations.

5.3.1 Thinning the chain

Thinning is the process of systematically discarding part of a chain to make it easier to store and process. Typically, a chain is thinned by keeping every rth value or some suitable value of r. A useful way of assessing the impact of thinning is to use the formula for the standard error of the mean of a first-order autocorrelated sequence of simulations:

$$\frac{\sigma}{\sqrt{n}}\sqrt{\frac{1+\rho}{1-\rho}}$$

Suppose that we start with a chain of length 10,000 that has an autocorrelation of 0.8. The formula tells us that its standard error will be about 0.03σ. If the autocorrelation declines approximately geometrically, then the correlation between values 5 apart will be about $0.8^5 = 0.33$, and after being thinned by 5, the resulting chain of length 2,000 will have an approximate standard error of 0.0315σ. Thinning does reduce the precision, but the increase in the standard error is only of the order of 5% of its value. Similar calculations show that thinning by 10 would produce a chain of length 1,000 with an approximate standard error of 0.0352σ, an increase of 17% over the full chain. The storage requirements for a chain of 10,000 are not that great, so thinning would be of limited value, but if a problem required a chain of a million simulations, then thinning the chain might well be a reasonable tradeoff of storage against precision. To guide these calculations, we can picture the decline in correlation at different lags by using `mcmcac`, a wrapper for Stata's `ac` command, as illustrated in figure 5.6.

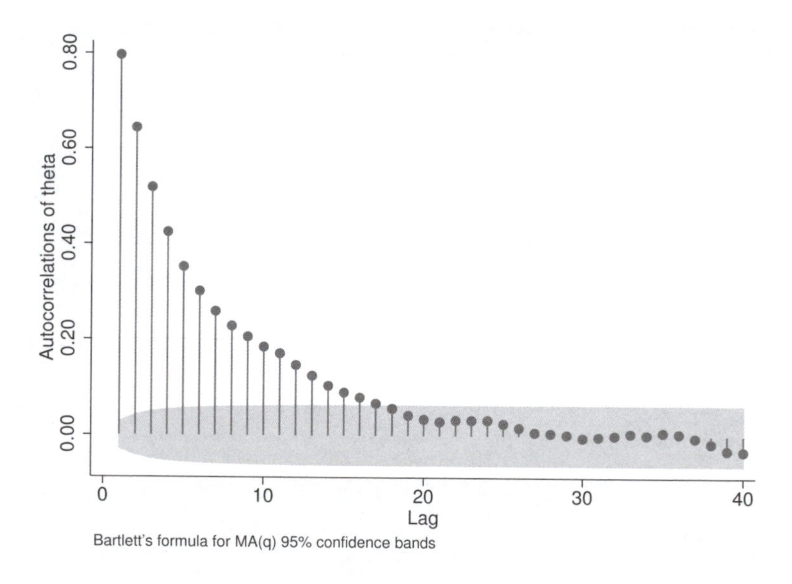

Figure 5.6. Autocorrelation within a chain of 1,000 observations

5.4 Running multiple chains

A good way to assess convergence is to run several chains from widely dispersed starting values to see when they come together. Typically, multiple chains will start from values, some of which are known to be extreme, so a longer than normal burn-in may be required. With multiple chains, it is a good idea to monitor the whole chain from its very beginning so that one can watch as the chains converge.

Figure 5.7 shows a plot for a parameter, θ, produced by `mcmcintervals` for three parallel chains simulated by using code similar to that described in section 5.2. The true value of the parameter is 5, but one chain starts with $\theta = 5$, another chain starts with $\theta = 8$ with linear drift for the first 500 iterations, and a third chain starts with $\theta = 1$ and drift for the first 250 iterations. The plot shows the means and 80% intervals for 10 successive sections of the iterations. It is clear from the plot that the 3 chains come into good agreement after about 500–600 iterations. In this case, the initial 600 iterations need to be discarded, and the remaining 400 iterations from each chain could be combined into a single set of 1,200 iterations for estimation. Whether 1,200 is enough depends on the property of the distribution to be captured and on the use to which it is put.

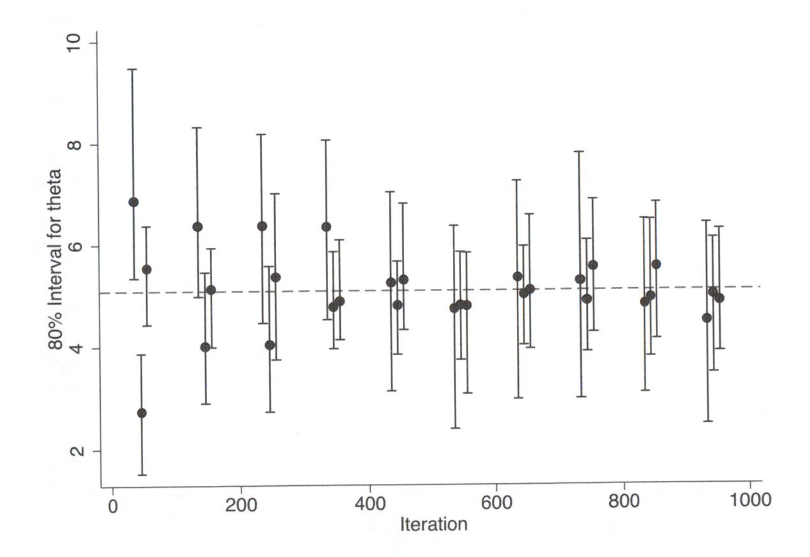

Figure 5.7. Interval plot of simulations from chains with high autocorrelation and different starting values

Brooks and Gelman (1998) and Gelman and Rubin (1992) investigated the use of a plot that compares the variability of the pooled chains with the average variability across the individual chains in a manner analogous to an analysis of variance. At convergence, these two measures of variability should become equal. Initially, the pooled and average variances were used, but unless the posterior distribution is approximately normal, it is probably better to use a nonparametric measure of variability such as the width of the interval between the 10% and 90% centiles. In that case, a simple measure of performance is

$$R = \frac{\text{interval from all the chains combined}}{\text{average of the intervals of the separate chains}}$$

For a subchain consisting of the first N values, R is calculated from the second half of the subchain and then plotted against N.

Figure 5.8 shows the Brooks–Gelman–Rubin plot corresponding to figure 5.7 and was created using `mcmcbgr`. The ratio, R, starts large but approaches 1 after about 600 iterations. In the same way, the two lines in the lower plot come together after about 600 iterations, thus leading to the same conclusions as we drew from the intervals plot.

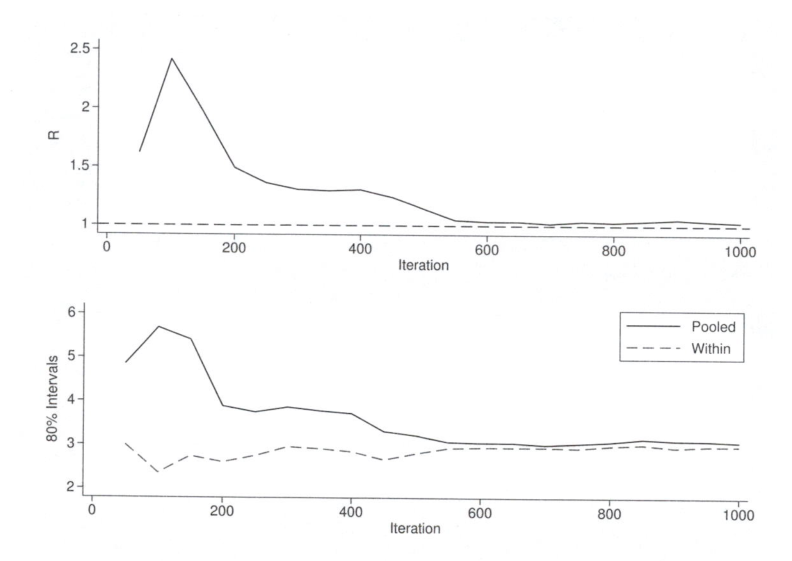

Figure 5.8. Brooks–Gelman–Rubin plot from the data displayed in figure 5.7

The Brooks–Gelman–Rubin plot is useful for investigating the impact of initial values on the estimation of the posterior and for detecting a multimodal solution. Its use is analogous to running a maximum likelihood analysis from different starting values to ensure that the algorithm does not settle on a local maximum.

5.5 Convergence of functions of the parameters

Because our real interest is in the convergence of the joint posterior distribution, inspection of the marginal distributions will tell only part of the story, and sometimes, functions that combine information on all, or on subsets, of the parameters can be more informative.

One obvious multiparameter function to use is the value of the joint log posterior, which could be calculated at the end of each full cycle of the MCMC algorithm. During the burn-in, the log posterior should move toward its maximum, after which it should spend most of its time close to that maximum. All the plots used to investigate individual parameters could equally be applied to the log posterior.

Sometimes, it is important to ensure that some function of the parameters is well estimated because that function is of special interest. For instance, when fitting a regression model with coefficients β_1 and β_2, we might be interested in the ratio of those parameters, in which case it would be sensible to calculate β_2/β_1 at each iteration to monitor the convergence of the ratio and to assess the size of the sampling error of the ratio.

A useful function for assessing general convergence is the distance between a vector of parameters simulated during one full iteration and the overall mean of the chain. To enable this to be monitored, the program `mcmcmahal` calculates and plots the Mahalanobis distance, a measure of distance that takes account of the correlation between parameters. The formula for the Mahalanobis distance is

$$D^2 = (\theta - \overline{\theta})'\Sigma^{-1}(\theta - \overline{\theta})$$

where $\overline{\theta}$ is the vector of parameter means, and Σ is the corresponding variance matrix. When the sample of simulations represents an independent sample from a log posterior that is quadratic, the distribution of distances will be chi-squared with degrees of freedom equal to the number of parameters. During the burn-in, this distance should get smaller, and after the burn-in, it should move backward and forward around a value close to the number of parameters. The use of `mcmcmahal` is demonstrated in case study 5.

5.6 Case study 5: Beta-blocker trials

Brophy, Joseph, and Rouleau (2001) reported a meta-analysis of 22 randomized, controlled trials of the use of beta-blockers for congestive heart failure in which the primary outcome was mortality. The results from the studies are shown in table 5.1. A Metropolis–Hastings algorithm will be used to fit a model to these data, but we will concentrate on the subsequent assessment of convergence.

Table 5.1. Mortality in the treated and placebo groups from 22 trials of beta-blockers for congestive heart failure

Study	beta-blocker	placebo
Anderson	5/25	6/25
Engelmeier	1/9	2/16
Pollock	0/12	0/7
Woodley	0/29	0/20
Paolisso	0/5	0/5
Waagstein	23/194	21/189
Wisenbaugh	1/11	0/13
Fisher	1/25	2/25
Bristow	4/105	2/34
CIBIS-I	53/320	67/321
Eichhorn	0/15	0/9
Metra	0/20	0/20
Olsen	1/36	0/23
Krum	3/33	2/16
Bristow	12/261	13/84
Packer	6/133	11/145
Colucci	2/232	5/134
Cohn	2/70	2/35
Aust/NZ	21/207	29/208
CIBIS-II	156/1327	228/1320
MERIT-HF	145/1990	217/2001
RESOLVD	8/214	17/212

The model will assume that the relative probability of dying between people on beta-blockers and on placebos, ϕ, is the same for all studies but that the baseline probability of dying in the placebo group varies from study to study. Thus, if p_i is the proportion dying on placebo in study i, the corresponding proportion dying on treatment will be ϕp_i. Potentially, this formulation could present problems if ϕp_i were to go over 1, but in practice, the probabilities for this problem are all small enough that this does not happen. An alternative formulation that avoids this potential problem is considered as an exercise at the end of the chapter.

Assuming that the trials and the patients within each trial are independent, the likelihood under this formulation will be the product of two binomials:

$$p(y|\phi, p_i) = \prod_{i=1}^{22} \left\{ \binom{n_{0i}}{y_{0i}} p_i^{y_{0i}} (1 - p_i)^{n_{0i} - y_{0i}} \right\} \left\{ \binom{n_{1i}}{y_{1i}} (\phi p_i)^{y_{1i}} (1 - \phi p_i)^{n_{1i} - y_{1i}} \right\}$$

Priors could be placed directly on the p_i to reflect our knowledge about the mortality rates in each study population, but often we will not have such detailed knowledge and instead will choose to treat all studies as having the same prior. This common prior could either be completely specified, for example, beta$(1, 5)$, or be made part of a hierarchical structure, such as beta(α, β), with higher-level priors placed on the parameters α and β so that we learn about their values from the data.

In this analysis, we will assume a common beta$(4, 36)$ distribution for all the p_i. This distribution has a mean of 0.1, so prior to seeing the data, we expect about 10% mortality in each of the placebo groups. The relative risk, ϕ, must be positive, and an ineffective treatment would be expected to have a value of 1, so a reasonable prior is gamma$(2, 0.5)$. These two priors are shown in figure 5.9.

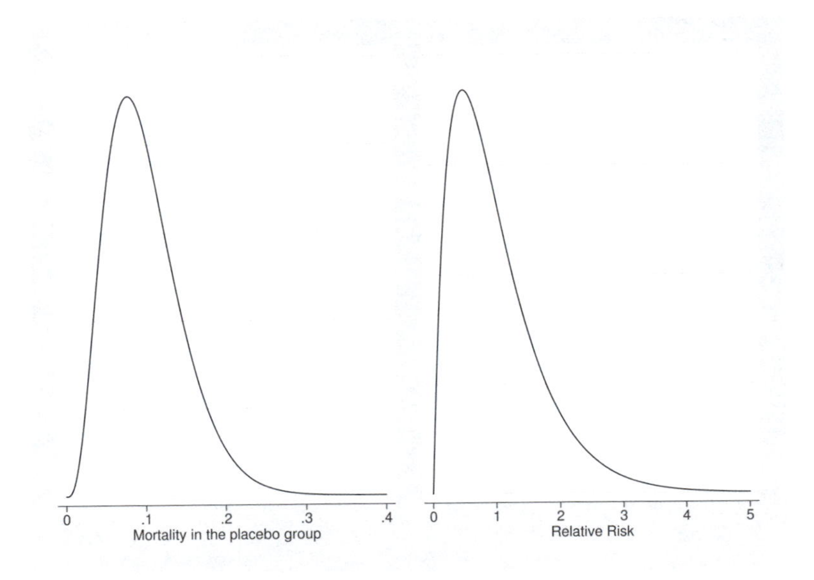

Figure 5.9. Prior distributions for the beta-blocker meta-analysis

Using the formulas for the likelihood and priors and ignoring unnecessary constants, we see that the joint posterior of all 23 parameters has the following form:

$$p(y|\phi, p_i) \propto \phi \exp(-2\phi) \prod_{i=1}^{22} p_i^3(1-p_i)^{35} \left\{p_i^{y_{0i}}(1-p_i)^{n_{0i}-y_{0i}}\right\} \left\{(\phi p_i)^{y_{1i}}(1-\phi p_i)^{n_{1i}-y_{1i}}\right\}$$

We can evaluate the conditional distributions efficiently using code that treats parameter 23, ϕ, differently from the other parameters.

begin: log posterior

```
program logpost
  args logp b p

  local phi = `b´[1,23]
  if `p´ == 23 {
     scalar `logp´ = log(`phi´) - 2*`phi´
     forvalues i=1/22 {
        local p_i = `b´[1,`i´]
        scalar `logp´ = `logp´ + (35+n0[`i´]-y0[`i´])*log(1-`p_i´) +          ///
             (y0[`i´]+3)*log(`p_i´) + (n1[`i´]-y1[`i´])*log(1-`phi´*`p_i´)     ///
             + y1[`i´]*log(`phi´*`p_i´)
     }
  }
  else {
     local p_i = `b´[1,`p´]
     scalar `logp´ = (35+n0[`p´]-y0[`p´])*log(1-`p_i´) +                       ///
        (y0[`p´]+3)*log(`p_i´)  + (n1[`p´]-y1[`p´])*log(1-`phi´*`p_i´) +       ///
        y1[`p´]*log(`phi´*`p_i´)
  }
end
```

end: log posterior

The code for fitting this model uses **mhstrnc** for the probabilities that must lie between 0 and 1 and **mhslogn** for the relative risk that must be positive. In the burn-in of 1,000 simulations, the standard deviations of the proposal distributions, **sigma**, are adapted to bring the acceptance rate close to 50%. These standard deviations are then fixed during the run of 25,000 simulations, from which every 5th is stored. The task of listing the samplers in **mcmcrun** is simplified because the first 22 parameters all have the same form and because the syntax allows us to place a number outside the bracket describing the sampler to indicate that it applies to more than one parameter.

begin: model fitting

```
matrix theta = J(1,23,.1)
matrix theta[1,23] = 1
matrix sigma = J(1,23,0.1)
mcmcrun logpost theta using temp.csv,                                  ///
    samp(22(mhstrnc, sd(sigma) lb(0) ub(1)) (mhslogn, sd(sigma)))  ///
    burn(1000) adapt update(25000) savelogp thin(5)                 ///
    par(p1-p22 phi) replace dots(0)
```

end: model fitting

The chain was run as shown above, and the stored values were checked for convergence. Trace and cusum plots were made of individual parameters, including **phi** and selected probabilities, with a focus on those with extreme data (not shown). However, because the number of parameters is moderately large, summary measures were also used to help assess the convergence. The value of the conditional log posterior calculated after updating **phi** was stored with the parameters, and its trace plot is shown in figure 5.10.

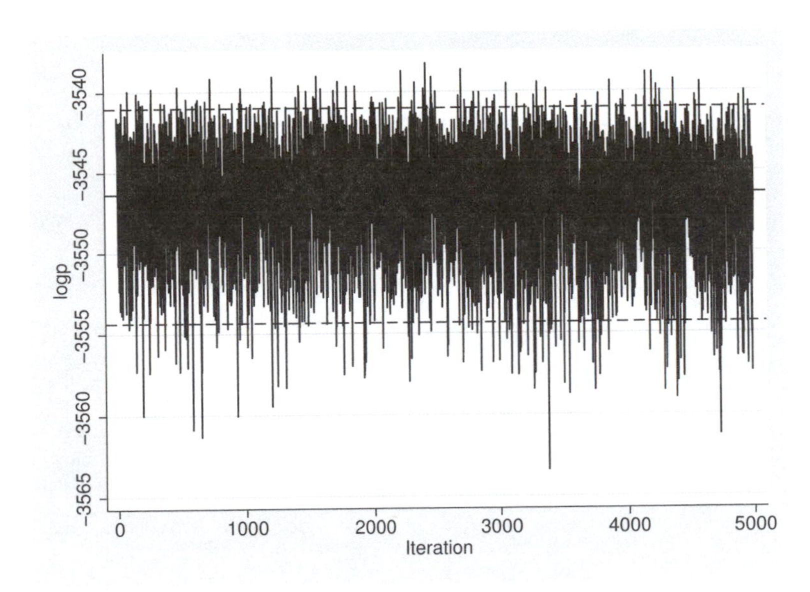

Figure 5.10. Trace plot of the value of the conditional log posterior for ϕ

As one would expect, there is no suggestion of drift, and the values stay close to the peak. The command `mcmcmahal` was used with the 22 probabilities to produce figure 5.11. Again there is no evidence of drift, and the chain seems well behaved. Indeed, the probability plot shows good agreement with the chi-squared distribution, which suggests that the log posterior must be approximately quadratic near its peak.

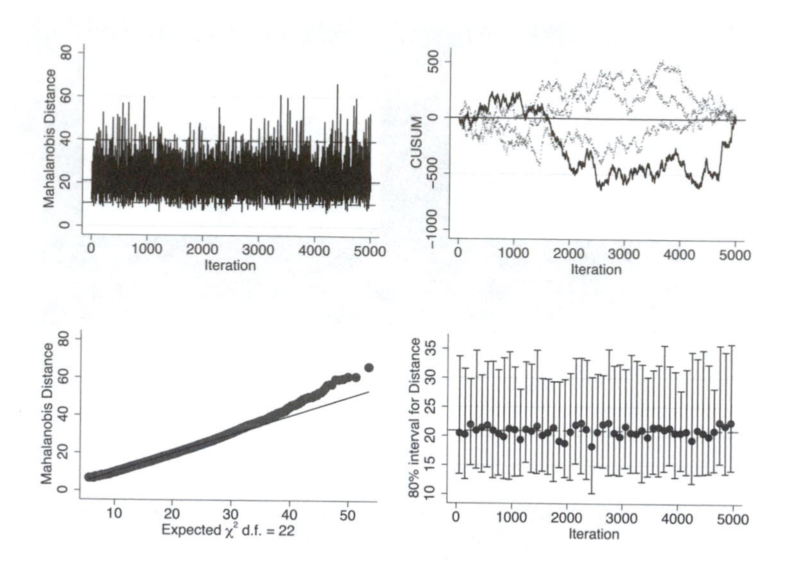

Figure 5.11. Summary plot of Mahalanobis distances for 5,000 simulations of the 22 baseline probabilities

5.7 Further reading

The most quoted review of methods for assessing convergence is Cowles and Carlin (1996), and that article contains references to most of the important work available up to that date. Brooks and Roberts (1998) provide a slightly more mathematical review of the same material, and El Adlouni, Favre, and Bobée (2006) give a practical comparison of convergence methods by testing them on three examples. Toft et al. (2007) and Sinharay (2004) both describe their experiences of applying convergence assessment techniques to specific examples. The software packages CODA, described by Plummer et al. (2006), and boa, described by Smith (2007), are both available in R and incorporate versions of many of the methods described in this chapter.

5.8 Exercises

1. Consider again the meta-analysis of beta-blockers described in section 5.6, but assume that instead of the relative probability, ϕ, being common across studies, it is the odds ratio, θ, that is common. If the probability of death in the placebo group is p_i and the probability in the treated group is q_i, then

$$\theta = \frac{q_i}{(1-q_i)} \frac{(1-p_i)}{p_i}$$

so that

$$q_i = \frac{\theta p_i}{1 - p_i + \theta p_i}$$

In section 5.6, a beta prior was placed on the level of mortality in the placebo groups. Some of the placebo groups were small, and so for them, this prior would have been influential. Rather than relying on our subjective prior, we might have used the baseline mortality in the other studies to inform our estimates for the small studies. Do this by creating a hierarchical model in which the logit of the mortality in the placebo groups follows a normal distribution with mean μ and precision τ. Set your own normal prior for μ and gamma prior for τ so that the conditional posteriors for those two parameters are conjugate.

Fit this hierarchical model with a run length sufficient to estimate the posterior mean and credible interval of θ to two decimal places of accuracy, then check the convergence.

2. Toft et al. (2007) describe their experience of assessing the convergence of the Hui–Walter model for evaluating diagnostic tests. Suppose that an imperfect test were to be compared with a gold standard to assess whether someone has a disease. The probabilities of the different possible outcomes might be represented by

		Gold Standard	
		−	+
Test	−	a	b
	+	c	d

Thus the probability that the test is positive and the person really has the disease is d. These probabilities could be reparameterized in terms of the sensitivity, $\text{Se} = d/(b+d)$; the specificity, $\text{Sp} = a/(a+c)$; and the disease prevalence, $p = b+d$.

In the absence of a Gold standard, we might instead choose to run two independent imperfect tests in two different populations. In this case, it might be reasonable to assume that the sensitivity and specificity vary by test but are the same in each population and that the disease prevalence is different in the two populations but the same for both tests. Adding a further assumption that the tests are independent given the true disease status enables us to calculate probabilities such as (++) the chance that both imperfect tests are positive, which will be $p\text{Se}_1\text{Se}_2 + (1-p)(1-\text{Sp}_1)(1-\text{Sp}_2)$. Hypothetical results of such an experiment in which 1,000 people from each population were subjected to each imperfect test were analyzed by Toft et al. (2007) and are shown below.

	++	+−	−+	−−
Population A	240	240	240	280
Population B	280	240	240	240

In population A, 280 people were disease free according to both of the imperfect tests. Toft et al. (2007) used flat beta distributions, $B(1,1)$, as their priors on the sensitivities and specificities of the two tests and the true prevalences in the two populations. There are in fact mirror solutions for this problem such that for any solution with a particular value, s, of a sensitivity, there will be an equivalent log likelihood for a solution with sensitivity $(1 - s)$. To avoid flipping between these solutions, place $U(0,5,1)$ priors on the sensitivities and specificities. Create a program for calculating the log posterior in terms of the six parameters. Fit the model, and assess convergence comparing your results with those of Toft et al. (2007).

6 Validating the Stata code and summarizing the results

6.1 Introduction

When one uses a standard Stata command, it is safe to assume that the underlying code has been thoroughly tested, first by StataCorp and then by the wider community of Stata users. In contrast, programs written for personal use are unlikely to be checked as thoroughly and so must be used with much more caution. In particular, user-written Bayesian analysis programs offer a lot of scope for error, most frequently because of mistakes in the algebra or in the coding of the program for evaluating the log posterior. For important analyses, it makes sense to validate such programs before going public with the results.

Once our program has been validated and run to convergence, the final step in the Bayesian analysis is the summary of the large sample of values generated from the joint posterior distribution. In a high-dimensional model, this joint distribution can be difficult to visualize, and so it is usual to concentrate on the marginal distributions of the individual parameters or on other low-dimensional derived quantities. Marginalizing any one parameter requires averaging (integrating) over all other parameters, but with a chain of simulations, this integration is trivial because it merely involves extracting the simulations for the parameter of interest and ignoring the others. Despite this simplicity, it is still important to remember that correlations between parameters will exist, and if one parameter were conditioned on particular values of the others, our conclusions about that parameter might well be different. If this is the case, then it is important that those dependencies are made clear.

To illustrate some of the methods for validating software and summarizing the results of a Bayesian analysis, we will fit an ordinal regression model to the data from a study previously considered by Ryu (2009).

6.2 Case study 6: Ordinal regression

The data in table 6.1 come from a colorectal cancer trial in which patients were randomized to one of three treatments: IFL, FOLFOX, or IROX. At their third treatment visit, the patients were asked to rate their appetite on an ordered five-point scale.

Table 6.1. Number of patients rating their appetite at each point of the scale

	Normal	Not always good	Don't really enjoy	Forced to eat	Can't stand
IFL	50	9	4	3	0
FOLFOX	24	13	14	2	0
IROX	31	14	8	2	0

Ryu (2009) used a standard cumulative logistic model for these data but was specially interested in one particular measure of the difference between treatment groups. This measure captures the chance that a randomly selected person on one treatment reports a lower score than someone on an alternative treatment. If the score on treatment k_1 is denoted by Y_{k_1}, the measure for comparing treatment k_1 with k_2 is defined as follows:

$$\theta_{k_1 k_2} = P(Y_{k_1} < Y_{k_2}) + \frac{1}{2} P(Y_{k_1} = Y_{k_2})$$

This measure is closely related to the Mann–Whitney statistic with an adjustment for ties.

The cumulative logistic model can be written as

$$\text{logit}(\pi_{jk}) = \alpha_j + \beta_k$$

where $\pi_{jk} = \Pr(Y_k \leq j)$ represents the probability of response of j or less in treatment group k, $j = 1, 2, 3, 4$, $k = 1, 2, 3$, and of course $\pi_{5k} = 1$. Here the β's measure the impact of treatment and require a constraint such as $\beta_1 = 0$, while the α's capture the increasing cumulative probabilities across the response categories and are constrained such that $\alpha_1 \leq \alpha_2 \leq \alpha_3 \leq \alpha_4$.

The analysis considered by Ryu (2009) had this same structure but was not Bayesian, so we will need to specify our own priors. The logit of an event that occurs 90% of the time is about 2, and the logit of one that occurs 10% of the time will be about -2, so we might confidently expect that α_1 will be between ± 2. Indeed, if we felt sure that more than half the people on IFL would grade their appetite as "normal", then we might restrict the range of α_1 to be positive. Capturing the ordering in the α's is easier if we reparameterize the model in terms of $\alpha_1, \delta_2, \delta_3, \delta_4$, where

$$\alpha_j = \alpha_{j-1} + \delta_j \qquad j = 2, 3, 4$$

Now to guarantee the ordering of the α's, the priors only have to ensure that all the δ's are positive; we will use a gamma distribution with a mean of 1 for each of the δ's. The treatment differences are unlikely to be huge, so we will use a fairly tight normal prior centered on zero for the β's. In summary, the priors to be used in our analysis are

$$\begin{aligned}
\alpha_1 &\sim N(0,1) \\
\delta_j &\sim G(5,0.2) \qquad j = 2,3,4 \\
\beta_k &\sim N(0,0.5) \qquad k = 2,3
\end{aligned}$$

We can see from the data that we will have problems estimating δ_4 because there were no observations in the final category of appetite; however, this should not influence our choice of priors, which, as always, must be stated without reference to the actual data.

Because the measures of treatment effect, θ, are of special interest, it makes sense to investigate the priors on θ implied by the priors that have been placed on the other parameters. In this case, the treatment comparisons 12 and 13 have the same prior, so it is sufficient to look at the implied prior on either of the θ's. This can be done by simulating a large set of data from the priors, calculating the probabilities for each cell of the 3×5 table, and then calculating each `theta`. The code used for doing this is shown below and produced the simulated distribution plotted in figure 6.1. The implicit prior on each `theta` has a mean of 0.5, which indicates no difference between treatments, and a standard deviation of about 0.065.

begin: simulating theta

```
set obs 10000
generate alpha1 = rnormal(0,1)
generate beta2  = rnormal(0,0.5)
generate beta3  = rnormal(0,0.5)
generate delta2 = rgamma(5,0.2)
generate delta3 = rgamma(5,0.2)
generate delta4 = rgamma(5,0.2)
generate alpha2 = alpha1 + delta2
generate alpha3 = alpha2 + delta3
generate alpha4 = alpha3 + delta4
generate beta1 = 0
forvalues k=1/3 {
  generate cp0 = 0
  forvalues j=1/4 {
local h = `j´ - 1
generate cp`j´ = invlogit(alpha`j´+beta`k´)
generate p`j´_`k´ = cp`j´ - cp`h´
}
  generate p5_`k´ = 1 - cp4
  drop cp*
}
generate theta12 = p1_1*(p2_2+p3_2+p4_2+p5_2)+ p2_1*(p3_2+p4_2+p5_2) ///
+ p3_1*(p4_2+p5_2) + p4_1*p5_2 + 0.5*(p1_1*p1_2 +    ///
p2_1*p2_2 + p3_1*p3_2 + p4_1*p4_2 +p5_1*p5_2)
generate theta13 = p1_1*(p2_3+p3_3+p4_3+p5_3)+ p2_1*(p3_3+p4_3+p5_3) ///
+ p3_1*(p4_3+p5_3) + p4_1*p5_3 + 0.5*(p1_1*p1_3 +    ///
p2_1*p2_3 + p3_1*p3_3 + p4_1*p4_3 +p5_1*p5_3)
generate theta23 = p1_2*(p2_3+p3_3+p4_3+p5_3)+ p2_2*(p3_3+p4_3+p5_3) ///
+ p3_2*(p4_3+p5_3) + p4_2*p5_3 + 0.5*(p1_2*p1_3 +    ///
p2_2*p2_3 + p3_2*p3_3 + p4_2*p4_3 +p5_2*p5_3)
```

end: simulating theta

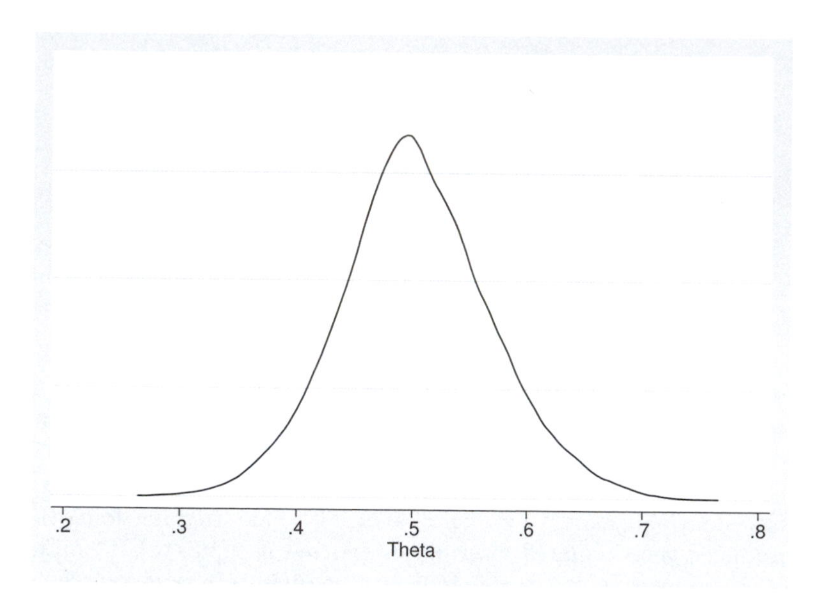

Figure 6.1. Implied prior on the measures of treatment difference, θ_{12} and θ_{13}

Fitting the model requires a program that evaluates the log posterior. For the code given below to work, the actual data from table 6.1 must be placed in a 3×5 matrix, C.

begin: ordinal regression

```
program logpost
   args logp b

   local alpha1 = `b´[1,1]
   local beta2  = `b´[1,2]
   local beta3  = `b´[1,3]
   local delta2 = `b´[1,4]
   local delta3 = `b´[1,5]
   local delta4 = `b´[1,6]
   local alpha2 = `alpha1´ + `delta2´
   local alpha3 = `alpha2´ + `delta3´
   local alpha4 = `alpha3´ + `delta4´
   local beta1 = 0
   local cp0 = 0
   forvalues k=1/3 {
      forvalues j=1/4 {
         local h = `j´ - 1
         local cp`j´ = invlogit(`alpha`j´´+`beta`k´´)
         local p`j´_`k´ = `cp`j´´ - `cp`h´´
      }
      local p5_`k´ = 1 - `cp4´
   }
   scalar `logp´ = 0
   forvalues k=1/3 {
      forvalues j=1/5 {
         scalar `logp´= `logp´ + C[`k´,`j´]*log(`p`j´_`k´´)
      }
   }
```

```
        logdensity normal `logp'  `alpha1' 0 1
        logdensity normal `logp'  `beta2'  0 0.5
        logdensity normal `logp'  `beta3'  0 0.5
        logdensity gamma  `logp'  `delta2' 5 0.2
        logdensity gamma  `logp'  `delta3' 5 0.2
        logdensity gamma  `logp'  `delta4' 5 0.2
end
```

end: ordinal regression

The model was fit using the Metropolis–Hastings algorithm with univariate normal or log-normal random-walk proposal distributions as described in chapter 3. During a burn-in of length 1,000, the standard deviations of the proposal distributions were tuned to give an approximate 50% acceptance rate. The subsequent run was of length 20,000, and every 4th value was saved, giving a stored chain of 5,000 sets of parameter values. Convergence was checked by using the methods of chapter 5. Marginal trace and cusum plots were inspected to look for any remaining influence of the initial values, and section plots were used to compare the first and second halves of the set of simulations. It is possible to investigate the posterior of the `theta`'s using the results from the Markov chain Monte Carlo (MCMC) run in much the same way as we did for the implicit prior on `theta`. Thus, from the simulated α's and β's, we can calculate the probabilities for each category of response in each treatment group, and from those probabilities, we can calculate the measures of difference between the two groups. We will return to the problem of summarizing the results for the α's, β's, and θ's in section 6.5, but first, we need to validate the code.

6.3 Validating the software

Creating the program for calculating the log posterior offers a lot of scope for error, both in the algebra and in the typing of the code, so it makes sense to check its accuracy before using the results. Sometimes, we can produce a likelihood analysis of the same model, thus enabling us to compare the likelihood and Bayesian results, perhaps after adjusting our priors to make them less informative. Alternatively, we might simulate some data and analyze them with the Bayesian program to see whether the estimates that it produces are close to the values used in the simulation.

Cook, Gelman, and Rubin (2006) described a simple method that extends the idea of analyzing simulated data. They noted that it is often comparatively easy to generate a random set of parameters from the priors; for instance, for the ordinal regression of case study 6, this could be done using Stata's `rnormal()` and `rgamma()` functions. Given a simulated set of parameters, it should be possible to generate a random dataset, and if those artificial data were analyzed using the Bayesian software, the estimates ought to resemble the parameter values used to generate the data. To confirm that this is the case, we could inspect the MCMC analysis parameter by parameter and calculate the quantiles, that is, the proportion of the MCMC chain that is less than the parameter value used in generating the data. If all is well, about 50% of the chain will be less than the true parameter value, and more generally the quantiles should be uniformly distributed between 0 and 1. So, to validate the software, we could repeat this whole process several times to see whether the quantiles for each parameter are indeed uniformly distributed.

Cook, Gelman, and Rubin (2006) suggested the construction of a chi-squared statistic to summarize such a set of replications, but there are many other summary measures that could be used instead.

The obvious drawback of Cook's method is the time that it takes to run. For the ordinal regression problem, the creation of one chain of length 20,000 for a single dataset takes about two minutes on a desktop computer, so analyzing many replicates will not be a trivial undertaking. However, it is better to spend an hour or two validating the program than to publish wrong results. Experience shows that really gross errors will be obvious within the first few replications and that more subtle problems will suggest themselves in, say, 20 simulated replications.

A useful spin-off of generating multiple sets of artificial data using parameters drawn from the priors is that they help to show the connection between the choice of priors and the data that those priors imply. This process can be a great help when asking experts to specify their prior beliefs. They can be asked for their priors and then be shown a few datasets simulated under those priors, and this may well cause them to reassess their priors. The process can act as a particularly useful lesson for people who say that they have extremely vague priors. If, despite seeing the consequences of their choices, the experts still hold to vague priors, it may be better to use narrower priors when validating the software than in the actual analysis.

For the ordinal regression problem, sets of parameters and artificial data can be created using

begin: artificial data

```
quietly set obs 174
generate trt = 1 + (_n>66) + (_n>119)
generate u = runiform()
generate appetite = 1
local alpha1 = rnormal(0,1)
local beta2  = rnormal(0,0.5)
local beta3  = rnormal(0,0.5)
local delta2 = rgamma(5,0.2)
local delta3 = rgamma(5,0.2)
local delta4 = rgamma(5,0.2)
local alpha2 = `alpha1´ + `delta2´
local alpha3 = `alpha2´ + `delta3´
local alpha4 = `alpha3´ + `delta4´
local beta1 = 0
forvalues k=1/3 {
  forvalues j=1/4 {
   local cp`j´ = invlogit(`alpha`j´´+`beta`k´´)
   quietly replace appetite = appetite + 1 if u > `cp`j´´ & trt == `k´
  }
}
matrix C = J(3,5,0)
forvalues i=1/3 {
   forvalues j=1/5 {
  quietly count if trt == `i´ & appetite == `j´
   matrix C[`i´,`j´] = r(N)
   }
}
```

end: artificial data

This code can be placed in a loop together with the code for the MCMC analysis and repeated 20 times with the estimated quantiles for each parameter posted to a results file. When saving the results, one should include each parameter's posterior mean, standard deviation, and standard error in the file because these can be helpful in diagnosing the cause of any problems. This process is illustrated in the following code, in which to save space, dots have replaced the commands that have already been given for generating the data and calculating the log posterior. Because simulated data are analyzed, we need to be even more conscious of the need for numerical precision in the calculation of the log posterior. With an extreme dataset, the estimate of `delta4` could become very large, in which case, some of the cell probabilities might become indistinguishable from zero, and the likelihood would be undefined. Following the recommendations of section 2.6, we add a line at the end of `logpost` to replace missing values by a large negative number. The effect will be to steer the Metropolis–Hastings algorithm away from such extreme regions of the parameter space.

begin: quantiles

```
program logpost
   args logp b

   local alpha1 = `b´[1,1]
...
   if missing(`logp´)  scalar `logp´ = -9E10
end

tempname rf
postfile `rf´ alpha1 q1 m1 sd1 se1 beta2  q2 m2 sd2 se2 ///
  beta3  q3 m3 sd3 se3 delta2 q4 m4 sd4 se4 ///
  delta3 q5 m5 sd5 se5 delta4 q6 m6 sd6 se6 ///
  using ordinal_valid.dta,replace
set seed 639125
forvalues rep=1/20 {
   quietly drop _all
   quietly set obs 174
   generate trt = 1 + (_n>66) + (_n>119)
...
   matrix b = (`alpha1´,`beta2´,`beta3´,`delta2´,`delta3´,`delta4´)
   matrix sd = J(1,6,0.5)
   mcmcrun logpost b using temp.csv,                  ///
      samp(3(mhsnorm, sd(sd)) 3(mhslogn, sd(sd)))     ///
      burnin(1000) adapt updates(20000) thin(4)       ///
      replace par(alpha1 beta2 beta3 delta2-delta4) dots(0) jpost
   import delimited temp.csv, clear
   local n = _N
   local postlist ""
   foreach v of varlist a* b* d* {
      quietly count if `v´ < ``v´´
      local q`v´ = (r(N)+0.5)/(`n´+1)
      quietly mcmcstats `v´
      local postlist "`postlist´ (``v´´) (`q`v´´)"
      local postlist "`postlist´ (`r(mn1)´) (`r(sd1)´) (`r(se1)´)"
   }
   post `rf´ `postlist´
}
postclose `rf´
```

```
use ordinal_valid.dta, clear
local i = 0
foreach v of varlist q* {
   local ++i
   generate z`i´ = invnormal(q`i´)
   generate c2`i´ = z`i´*z`i´
   quietly summarize c2`i´
   display %6s "`v´" %8.3f `r(sum)´ %8.4f chi2tail(_N,`r(sum)´)
}
```
end: quantiles

For illustration, consider alpha1. Twenty replications of the ordinal regression code gave the following quantiles:

```
0.3902  0.7093  0.7581  0.1485  0.1066
0.4352  0.8085  0.7955  0.3259  0.5170
0.4074  0.3451  0.6653  0.6545  0.1485
0.3485  0.5008  0.4760  0.0940  0.8473
```

Cook, Gelman, and Rubin (2006) suggested a chi-squared statistic for combining such quantiles,

$$X^2 = \sum_{i=1}^{20} \left\{\Phi^{-1}(q_i)\right\}^2$$

where q_i is the quantile, and Φ is the cumulative normal function. If the quantiles are uniformly distributed, this statistic will be chi-squared with 20 degrees of freedom. For alpha1, the value of this statistic is 21.01, which has a p-value of 0.40, and in this example, the six parameters have p-values of 0.40, 0.53, 0.79, 0.41, 0.36, and 0.72, which suggests that our program is performing correctly.

To demonstrate the ability of the method to detect an error, we modified the program for calculating the log posterior so that the beta's are subtracted rather than added to the alpha's.

```
local cp`j´ = invlogit(`alpha`j´´-`beta`k´´)
```

When a simulated beta is small, the error is hard to detect, but when a large beta is drawn from the prior, it becomes obvious from the quantiles that there is a problem with the program. In this case, there is hardly any need to run all 20 replicates, but for illustration, we did this, and the p-values for the six parameters using Cook's statistic were 0.41, 0.00, 0.00, 0.42, 0.37, and 0.74. The quantiles point to a problem with parameters 2 and 3, beta2 and beta3.

6.4 Numerical summaries

Once we are confident that our code is correct, we can summarize and explore the MCMC chain. The program mcmcstats provides a convenient numerical summary of a set of MCMC simulations. The output that it produces for the ordinal regression analysis is

```
. mcmcstats alpha1 be* de*
------------------------------------------------------------------------------
Parameter          n       mean       sd        sem    median          95% CrI
------------------------------------------------------------------------------
 alpha1         5000      0.749    0.209    0.0053     0.748 (    0.353,    1.165 )
 beta2          5000     -0.712    0.280    0.0062    -0.706 (   -1.274,   -0.186 )
 beta3          5000     -0.328    0.288    0.0063    -0.333 (   -0.883,    0.255 )
 delta2         5000      1.071    0.157    0.0028     1.065 (    0.782,    1.393 )
 delta3         5000      1.566    0.288    0.0049     1.550 (    1.055,    2.167 )
 delta4         5000      1.617    0.483    0.0082     1.566 (    0.834,    2.696 )
------------------------------------------------------------------------------
```

The first three columns of the output give the size, mean, and standard deviation of the chain. Next comes `sem`, the standard error of the mean, which is calculated using Stata's `prais` command to make an adjustment for the autocorrelation. We can see that in this case, if the simulations had been independent, the standard error of `alpha1` would have been $0.209/\sqrt{5000} = 0.0030$, so the autocorrelation in the chain has roughly doubled the standard error. The value of `sem` describes the sampling accuracy of the mean. Here the mean of the posterior of `alpha1` is accurate to about 2 standard errors or ± 0.011. If this is not sufficient for our purposes, then we should run a longer chain.

Theoretically, the choice of the best single value to summarize a posterior distribution is justified by placing a loss function over the distance between the point estimate and the true value; a quadratic function leads us to use the mean, and a loss function proportional to the absolute difference leads to the median. In practice, the choice is usually much less formal. Most Bayesians use the mean of the posterior as the natural point estimate for a parameter unless the posterior is very skewed, in which case they might use the median.

The 95% credible intervals (CrI) play a similar role to 95% confidence intervals, except that their interpretation is more straightforward and arguably more relevant. The CrI for `beta2`, the parameter that measures the difference between the first two treatments, is $(-1.274, -0.186)$, so 95% of the posterior distribution lies within that range, and we can say that we believe with 95% probability that `beta1` lies between those limits. Here, because zero is outside the interval, there is a high probability that scores for IFL are generally lower than those for FOLFOX. This is the way that many people interpret confidence intervals, although of course they should not.

A 95% credible interval is obtained by excluding 2.5% of the posterior distribution in each tail. One argument against credible intervals is that they may exclude values in one tail that have a higher posterior density than some included values from the other tail; thus some Bayesians prefer to create a highest posterior density (HPD) interval. This interval also includes 95% of the distribution but does so by ensuring that at the two ends of the range, the height of the posterior is the same. Besides ensuring that all values within the HPD interval have higher posterior density than those that are excluded, this also produces the shortest possible interval. Both credible and HPD intervals can be calculated by `mcmcstats`, although the more commonly used credible intervals are the default. The two types of intervals are illustrated in figure 6.2; both have the same total shaded area but in the credible interval on the left, the shaded area is equally divided between the two tails, while in the HPD interval, the two limits have equal posterior density. For approximately symmetric posterior distributions, the difference between the two types of intervals is negligible.

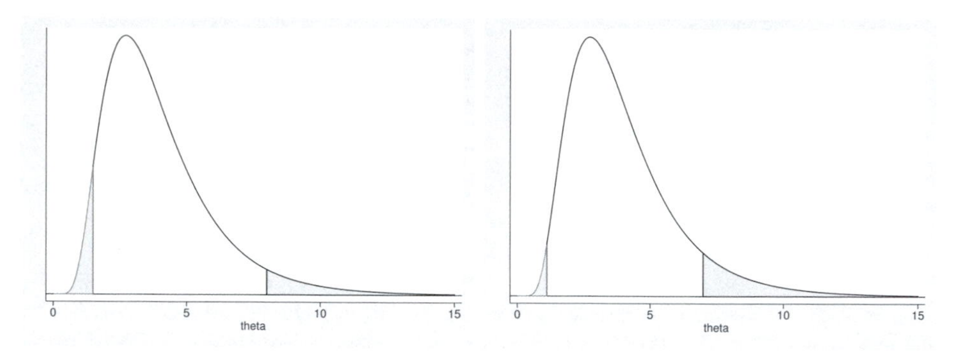

Figure 6.2. Diagrammatic representation of a credible interval (left) and a highest posterior density interval (right)

The `correlation` option of `mcmcstats` will add the matrix of correlations between the parameters to the output. This has two primary uses: first, it will tell us whether our beliefs about any of the parameters are dependent on other parameters, which may be important for interpretation; and second, it will alert us to high correlations that might have affected the speed of convergence of the MCMC chain. As we saw in chapter 3, when parameters are highly correlated, algorithms that move one parameter at a time can be very inefficient, and a better algorithm can sometimes be designed either by updating the correlated parameters as a block or by reparameterizing the model. For this model, only the estimates of `alpha1` and the two `beta`'s are noticeably correlated. So when interpreting the treatment difference, we should note that our estimates will depend to some extent on what we believe about `alpha1`.

```
             |  alpha1     beta2     beta3    delta2    delta3    delta4
-------------+------------------------------------------------------------
      alpha1 |  1.0000
       beta2 | -0.5636    1.0000
       beta3 | -0.5624    0.3493    1.0000
      delta2 | -0.1307   -0.0933   -0.0371    1.0000
      delta3 | -0.0123   -0.0408   -0.0185   -0.0566    1.0000
      delta4 |  0.0011   -0.0071    0.0009    0.0088   -0.0165    1.0000
```

A further option available with `mcmcstats` will create a Stata data file containing the results summarized in the basic output table. This can be useful, for instance, when we wish to display the summary statistics graphically. In this example, we might have saved the output and then plotted the posterior means of the parameters with lines to show their credible intervals. The following commands will create figure 6.3:

```
mcmcstats alpha1 b* d*, save(ordinal_stats.dta,replace)
use ordinal_stats.dta, clear
generate t = _n
twoway (scatter mean t) (rcap lb ub t), leg(off) yline(0) xtitle("") ///
  xlabel(1 "alpha" 2 "beta2" 3 "beta3" 4 "delta2" 5 "delta3" 6 "delta4")
```

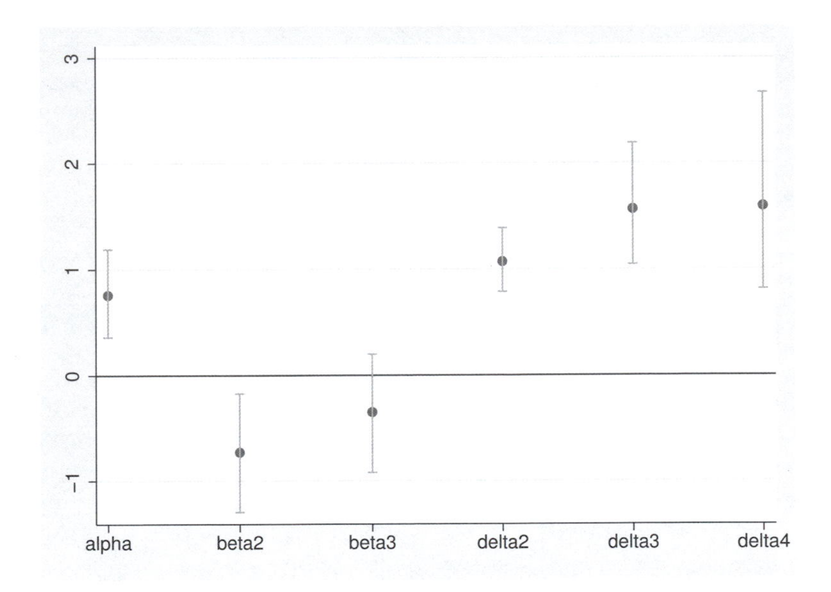

Figure 6.3. Posterior means and 95% credible intervals for the six parameters

Once the derived θ's have been calculated from each set of parameters in the chain, their summaries can also be obtained from `mcmcstats`.

```
. mcmcstats theta*
---------------------------------------------------------------------------
Parameter          n      mean        sd       sem    median          95% CrI
---------------------------------------------------------------------------
 theta12        5000     0.590     0.036    0.0007     0.590 (   0.523,    0.661 )
 theta13        5000     0.540     0.035    0.0007     0.540 (   0.470,    0.607 )
 theta23        5000     0.449     0.043    0.0007     0.448 (   0.365,    0.534 )
---------------------------------------------------------------------------
```

Ryu (2009) used maximum likelihood and found `theta12` to be 0.655 with a standard error of about 0.05; if this is averaged with the implicit prior on `theta12`, which had a mean of 0.5 and a standard deviation of 0.065, then we obtain a good approximation to the Bayesian estimate as $0.590 \approx (0.5/0.065^2 + 0.655/0.05^2)/(1/0.065^2 + 1/0.05^2)$ and $0.036 \approx 1/\sqrt{1/0.065^2 + 1/0.05^2}$. So the only differences between the results of Ryu (2009) and those obtained by the Bayesian analysis stem from the chosen priors. Placing vague priors on the model parameters would have made the implied priors on the `theta`'s less precise and would have made the Bayesian and likelihood solutions numerically similar.

Another useful feature of having an MCMC chain is that it enables us to compare parameters. For instance, we might wish to know whether IFL is more dissimilar to FOLFOX than IROX, $\theta_{12} > \theta_{13}$. Using the command

```
count if theta12 > theta13
```

tells us that about 88% of the simulations satisfy this inequality, so the probability that the difference between IFL and FOLFOX really is greater than the difference between IFL and IROX is 0.88. Alternatively, we could calculate `dtheta`, the difference between `theta12` and `theta13`, and summarize its values.

```
. generate dtheta = theta12 - theta13
. mcmcstats dtheta
----------------------------------------------------------------------------
Parameter            n      mean        sd       sem    median           95% CrI
----------------------------------------------------------------------------
 dtheta           5000     0.051     0.043    0.0007     0.051 (  -0.034,    0.135 )
----------------------------------------------------------------------------
```

As we have already seen, the maximum likelihood and Bayesian estimates of the `theta`'s differ because the chosen priors pulled the posterior distributions toward 0.5. To imitate a likelihood analysis, the priors must be made less informative. Relaxing the priors on `delta` makes convergence much slower without much effect on `theta`; however, if we widen the priors on `alpha1` and the `beta`'s, the implicit priors on `theta` are still centered on 0.5 but with a larger standard deviation. When $N(0, 3)$ priors are used for those parameters, the posterior means and 95% credible intervals for the `theta`'s become 0.65 (0.57,0.73), 0.59 (0.51,0.67), and 0.44 (0.34,0.53). Ryu (2009) calculated various approximate 95% confidence intervals, but for those based on a Wald approximation, the corresponding estimates were 0.66 (0.55,0.76), 0.59 (0.49,0.70), and 0.43 (0.31,0.55), numerically similar to this Bayesian analysis.

6.5 Graphical summaries

While the numerical values provided by `mcmcstats` are usually the best way of summarizing the posterior distribution, it is still good practice to plot the marginal densities to inspect the shapes of the densities. Figure 6.4 was created from the MCMC chain for the ordinal regression using the command `mcmcdensity`. It shows the symmetry of most of the distributions and the slight skewness of `delta4` brought about by our choice of prior and the lack of information about the most extreme category.

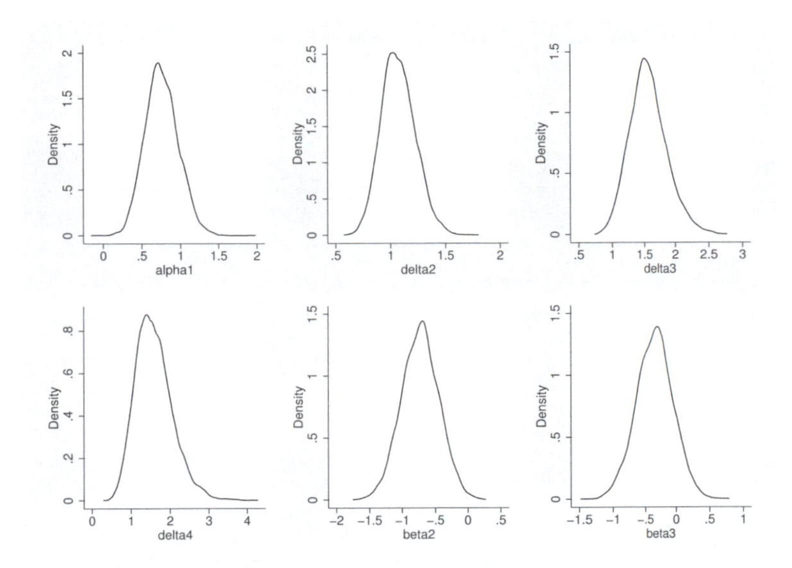

Figure 6.4. Marginal posterior densities drawn using `mcmcdensity`

The command `mcmcdensity` is a wrapper for Stata's `kdensity` command that plots smoothed univariate densities. It has the slight advantage of allowing for restricted ranges. If one knows that a parameter must be positive, then `mcmcdensity` can be told not to place any probability on negative values (see the help file for details).

We anticipated that `delta4` would be difficult to estimate because of the lack of any observations in the final category of appetite. It might be of interest, therefore, to see how the data have moved our prior beliefs about this parameter. The `addplot()` option of `mcmcdensity` allows us to superimpose the prior on the smoothed density, so figure 6.5 was produced by the following command:

```
. mcmcdensity delta4, addplot(function y=gammaden(5,0.2,0,x), range(0 4))
```

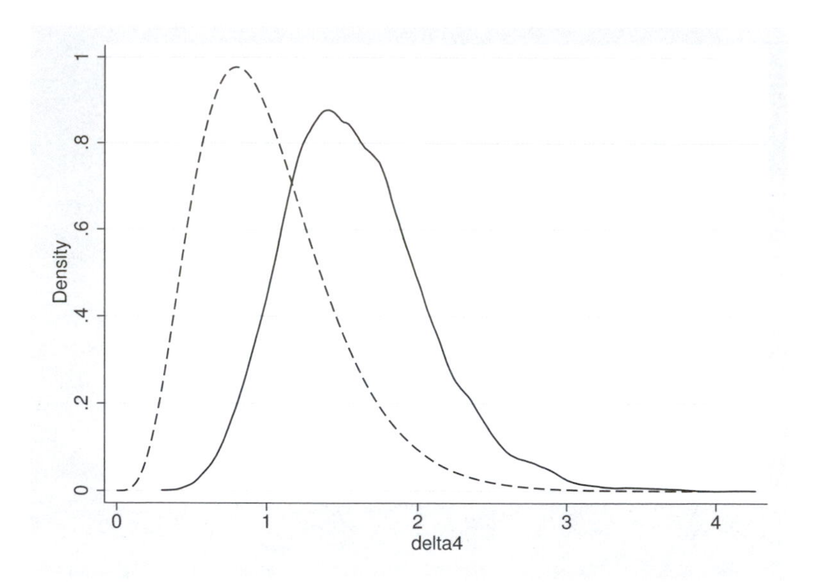

Figure 6.5. Posterior (solid) and prior (dashed) for `delta4`

The investigation of the relationships within sets of parameter estimates is generally not given the weight that it deserves, probably because it is difficult to do thoroughly in even moderately complex problems. We have seen that `mcmcstats` will, if requested, give the correlation matrix of a set of parameters, and this is the minimum summary that should be looked at. A graphical alternative to the correlation matrix is created by Stata's `graph matrix` command, as illustrated for the ordinal regression by figure 6.6.

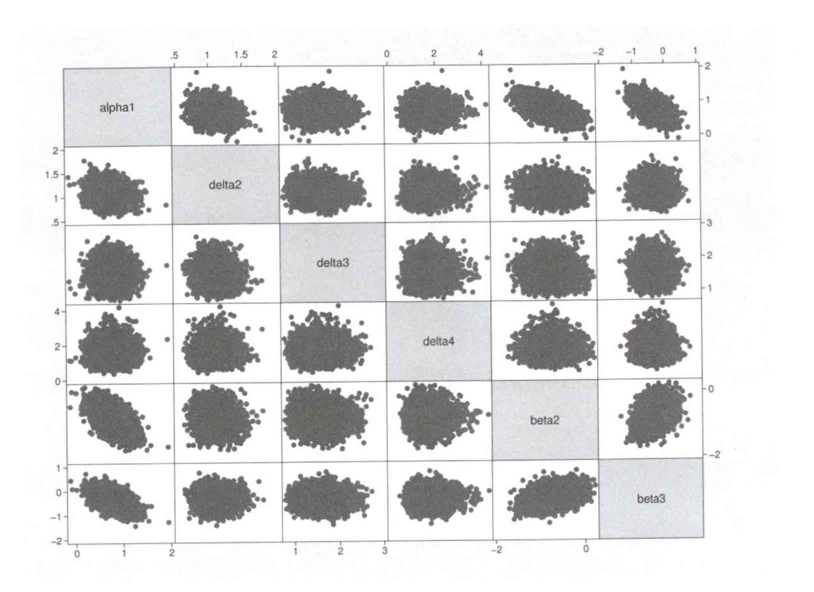

Figure 6.6. Matrix plot of the bivariate distributions of the ordinal regression parameters

In Stata 12, the `twoway contour` command was introduced for creating contour plots. These plots can be used to show a bivariate distribution. For instance, to show the joint distribution of `beta2` and `beta3`, we could use the `twoway contour` command. The contour plotting routine is slow when there are many points, so it makes sense to plot the curve using a sample of about 100 points taken from widely spaced points along the chain. In the code below, a normal distribution is used to smooth the simulated values, and then every 50th value is used for the contour plot. The choice of standard deviation for the normal kernels will influence the degree of smoothing, and the value of 0.1 used here was found by trial and error. The result is figure 6.7.

begin: contour plot

```
generate z = 0
forvalues i=1/5000 {
  local m1 = beta2[`i']
  local m2 = beta3[`i']
  generate tmp = normalden(beta2,`m1',0.1)*normalden(beta3,`m2',0.1)
  quietly summarize tmp
  quietly replace z = r(sum) in `i'
  drop tmp
}
sort beta2
twoway contour z beta3 beta2 if mod(_n,50) == 0, level(10) clegend(off) ///
   xscale(range(-1.4 0.2)) xlabel(-1.4(0.2)0.2)
```

end: contour plot

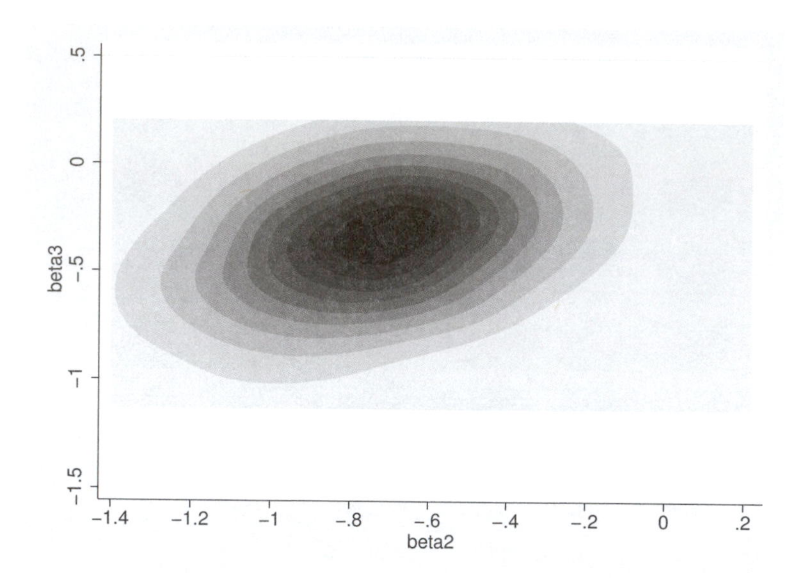

Figure 6.7. Contour plot of the joint posterior of `beta2` and `beta3`.

6.6 Further reading

This chapter has considered two areas of Bayesian analysis that have been largely ignored in the literature. Apart from the method of Cook, Gelman, and Rubin (2006) described in this chapter, the only other procedure for validating Bayesian software was described by Geweke (2004).

Wright (1986) and Chen and Shao (1999) discuss the calculation of HPD intervals. Surprisingly little work has been done on summarizing high-dimensional posterior distributions, and most people rely almost entirely on marginal summaries and plots; for exceptions, see Gelman (2004) and Kerman et al. (2008). There is a great need for data-mining techniques that can be applied to explore the high-dimensional MCMC chains to identify unusual patterns that are important either for improving convergence or for better understanding of the results. For instance, the method of projection pursuit, first suggested by Friedman and Tukey (1974), might be modified for exploring a multidimensional MCMC chain.

6.7 Exercises

1. The data below were analyzed using generalized estimating equations (GEEs) by Stiger, Barnhart, and Williamson (1999). They show the results of a clinical trial of a hypnotic drug versus placebo in people with insomnia. Before treatment and then again two weeks after treatment, the subjects were asked about the time it

took them to fall asleep. Their responses were categorized as under 20 minutes, 20–30 minutes, 30–60 minutes, and over an hour.

Table 6.2. Number of subjects reporting different lengths of time needed to fall asleep

	Baseline	Follow-up			
		<20	20–30	30–60	>60
Active	<20	7	4	1	0
	20–30	11	5	2	2
	30–60	13	23	3	1
	>60	9	17	13	8
Placebo	<20	7	4	2	1
	20–30	14	5	1	0
	30–60	6	9	18	2
	>60	4	11	14	22

a. Many models could be used with these data. For instance, Stiger uses a model that is parameterized in terms of a treatment effect, an occasion effect, and their interaction. However, we will treat the baseline state as a covariate rather than a response and fit

$$\text{logit}(\pi_{ijk}) = \alpha_i + \beta_j + \theta_k$$

where π_{ijk} represent the cumulative probabilities for baseline categories $i = 1, 2, 3, 4$, responses $j = 2, 3$, and treatment $k =$ placebo/active. Thus the α_i can be thought of as defining the chance of a follow-up response <20 for each baseline group. The β_j create the cumulative probabilities for the 20–30 and 30–60 responses, and the treatment effect θ distinguishes active from placebo. Set your own priors, and write a program that evaluates the log posterior.

b. Verify that your code is correct using the method of Cook, Gelman, and Rubin (2006).

c. Fit the model, and then select numerical and graphical summaries that convey the important properties of the joint posterior.

d. It is possible that the drug will have little effect on people who naturally fall asleep relatively quickly but will have a large effect on those who, without treatment, take a long time to fall asleep. Modify the model so that the treatment effect is different for different categories of baseline delay in falling asleep. Produce summaries that convey the difference between the joint posterior of this model and that of the previous model.

2. Geweke (2004) suggested a technique for validating a Bayesian analysis program. His method involves generating repeated samples of both parameters and associated data using two different approaches. The first approach, called the "marginal

conditional", works exactly as in Cook's method; that is, a set of parameter values is generated from the prior, and they are used to generate an artificial set of data. The second approach is called the "successive conditional" and starts like the marginal conditional with a sample of parameters from the prior and an artificial dataset, but it continues by running a step of the MCMC program with the artificial data to produce updated parameter values. The updated parameters are used to generate a new dataset, which is used to further update the parameters, which in turn generate a new dataset, and so on.

Each approach produces lots of sets (θ_i, y_i), where θ represents the parameters, and y represents the corresponding data. A statistic is chosen, such as the mean of θ, that the user feels will identify any errors in the program, and the chosen statistic is calculated using each of the sets generated by the two procedures. If the program is correct, then the two approaches should give equivalent summary statistics, so they can be compared in a Wald test.

a. Read Geweke's (2004) article, and then write a Stata program that implements the method for the ordinal regression analysis of the appetite data described by Ryu (2009). Run the program and assess the validity of the code.
b. Introduce an error into your program for calculating the log posterior, and investigate how well it is detected.
c. What are the comparative merits of the methods of Cook, Gelman, and Rubin (2006) and Geweke (2004)?

7 Bayesian analysis with Mata

7.1 Introduction

Using Stata's macro-substitution language to code a Markov chain Monte Carlo (MCMC) algorithm is a practical approach to the Bayesian analysis of a real problem provided that the number of parameters, the size of the dataset, and the length of the chain are all reasonably small. For larger problems, the speed of execution of the macro-substitution language becomes a major issue, and it can make Stata too slow for practical use. Fortunately, there are ways around this problem. In the next two chapters, we will review the alternative approaches: first, we will consider how Stata's built-in matrix language, Mata, can be used to speed up the calculations; and second, in the next chapter, we will see how to export a Bayesian analysis to WinBUGS or OpenBUGS, which are free, purpose-written programs for MCMC sampling that can be downloaded from the Internet and used in conjunction with Stata.

Mata is a C-like language that is fully integrated into Stata, so it can refer to Stata datasets and be run through the Stata interface. When deciding whether to use Mata or Stata, one must distinguish between the creation of the MCMC simulations themselves and the many other features of a Bayesian analysis, such as checking the convergence or summarizing the results. Tasks that are only performed a few times, for instance, plotting the trace of a parameter, are perfectly practical in Stata, but the MCMC calculations, which may need to be repeated hundreds of thousands of times, often require faster code.

All the samplers introduced in the previous chapters have Mata equivalents that are stored together in a downloadable library called `libmcmc`. This Mata code can be requested when using the `mcmcrun` command simply by adding the `mata` option. Thus all that the user needs to do is to write the program for evaluating the log posterior using Mata; then the analysis will proceed in the same way as it did with Stata, except that it will be faster by a factor of perhaps 10 or 20 times.

7.2 The basics of Mata

Stata commands are interpreted afresh each time they are used; that is, each line of a Stata program has its macros replaced by their current values, and then the resulting code is inspected and turned into a command that Stata can execute. For this reason, loops can be very slow because every line of code within the loop will be interpreted over and over again. Mata, on the other hand, is a compiled language, which means that its code is turned into executable commands once and only once before it is first used; thus the overhead associated with repeated interpretation is removed. For tasks that require the same operations to be repeated many thousands of times, the difference in speed between Stata and Mata can be quite dramatic.

Mata is not difficult to use, but its syntax is different from that of Stata, so there is an initial learning curve. This section gives an overview of the Mata code that is needed to run a simple MCMC analysis. Attention is deliberately restricted to the essentials needed for this specific task, and no attempt is made to be comprehensive. However, as our concern is with speed of execution, we will make our rather limited code as efficient as possible. In particular, this means using the matrix facilities of the language because these are very efficiently coded and because they are parallelized in versions of Stata that can use more than one processor.

To illustrate how simple it is to fit a model using Mata, we will consider a logistic regression in which the probability of y successes out of n trials is modeled in terms of a covariate x with quite vague normal priors on the regression coefficients. Specifically,

$$y_i \sim B(p_i, n_i)$$
$$\text{logit}(p_i) = \beta_0 + \beta_1 \text{x}_i$$
$$\beta_j \sim N(0,4) \quad j = 0,1$$

To fit this model using Stata, we need to write a program that evaluates the log posterior and then use `mcmcrun` to create the MCMC simulations using our choice of samplers. The code below generates some random data and then fits the logistic regression.

begin: logistic regression in Stata

```
program logpost
   args logp b ipar

   tempvar pr
   generate `pr´ = invlogit(`b´[1,1]+`b´[1,2]*x)
   scalar `logp´ = 0
   logdensity binomial `logp´ y n `pr´
   logdensity normal `logp´ `b´[1,1] 0 4
   logdensity normal `logp´ `b´[1,2] 0 4
end

set obs 50
set seed 407188
generate n = 1 + floor(runiform()*10)
generate x = rnormal(0,1)
generate p = invlogit(-1 + 0.5*x)
generate y = rbinomial(n,p)
matrix theta = J(1,2,0)
matrix s = J(1,2,0.2)
mcmcrun logpost theta using temp.csv, samp(2(mhsnorm, sd(s)))     ///
  burn(1000) adapt update(5000) jpost dots(0) savelogp            ///
  par(beta0 beta1)  replace
import delimited temp.csv, clear
```

end: logistic regression in Stata

The 6,000 iterations take 6.4 seconds using Stata/SE for Windows, and of that time, 87% is spent evaluating the log posterior with the function `logpost()`. Clearly anything that can speed up the calculation of the log posterior will greatly improve the overall performance.

To run exactly the same model using Mata requires very similar code:

begin: logistic regression in Mata

```
set matastrict off
mata:
mata clear
real scalar logpost(real matrix X,real rowvector b,real scalar ipar)
{
  p = invlogit(b[1]:+b[2]*X[,3])
  return(logdensity("binomial",X[,1],X[,2],p) + logdensity("normal",b,0,4))
}
end

set seed 407188
set obs 50
generate n = 1 + floor(runiform()*10)
generate x = rnormal(0,1)
generate p = invlogit(-1 + 0.5*x)
generate y = rbinomial(n,p)
matrix theta = J(1,3,0)
matrix s = J(1,3,0.2)
mcmcrun logpost X theta using temp.csv, samp(2(mhsnorm, sd(s))) ///
   burn(1000) adapt update(5000) jpost dots(0) savelogp        ///
   par(beta0 beta1)  replace data(X=(y n x) theta s) mata
import delimited temp.csv, clear
mcmcstats beta*
```

end: logistic regression in Mata

The most important difference between the Mata version and the previous Stata program is that the Mata version takes just 0.38 seconds to run, quicker by a factor of about 17 when compared with its Stata equivalent. To achieve this gain, we change the code in two places: first, we use slightly different options when calling `mcmcrun`; and second, we program the calculation of the log posterior using Mata.

When we set the `mata` option with `mcmcrun`, the command writes and runs a do-file containing the Mata code needed to run the analysis, including calls to our Mata function for calculating the log posterior and to the chosen samplers from `libmcmc`.

In Mata, the data needed to calculate the log posterior must be passed to the function as one of its arguments. To do this, we need to pack the data into a Mata matrix, and the name of that matrix must be given in the list of items that follows the command word `mcmcrun`, where it is placed between the name of the function and the name of the row vector of parameters. This was unnecessary when using Stata because all the variables currently in Stata's memory are available to the Stata calculations. This is not the case in Mata because Stata and Mata use different areas of memory, so when a Mata function is to access variables or matrices defined in Stata, the coding is simpler if they are first copied into Mata's memory. When the `mcmcrun` command is used, this transfer is performed automatically.

The data() option of mcmcrun identifies the variables or matrices to be transferred from Stata to Mata's memory. In this example, the Stata matrices beta and s are copied and stored as matrices in Mata under those same names, but the Stata variables are treated slightly differently. The three Stata variables y, n, and x that are listed in parentheses within the data() option are copied into Mata and stored in a single 50×3 matrix called X. When we want to refer to the data from the Stata variable n, we have to remember that it went into the second column of X and refer to it as X[,2] in our Mata programs.

In the example, we have used the nodots option with mcmcrun to suppress the printing of dots after every 250 iterations. These dots enable the user to monitor the progress of the analysis and make almost no difference to the run time when using Stata, but with Mata, the dots sometimes slow down the program noticeably (in this example, by a factor of about three times, although the exact gain will vary from computer to computer and will depend on the number of processors that Stata can use). The problem is that the Mata program has to wait every time that something is written to the screen. To avoid long delays, Mata's default action is to buffer lines of screen output and to display them when the line is completed. However, this is unhelpful if one wants to monitor progress; therefore, mcmcrun forces Mata to flush the display buffer after every dot. Because the calculations are so fast, the tiny delay during the display of the dots can have a noticeable impact on performance.

The Mata code can be prepared in the Do-file Editor and run in the same way as its Stata equivalent. The command mata: marks the start of a block of Mata code, which finishes with the command end. The code outside these limits is just standard Stata, so we will now concentrate on the section that calculates the log posterior.

The first thing that you will notice about the Mata version of logpost is that it is declared to be a real scalar. This means that logpost() is expected to return a scalar, in this case, one equal to the value of the log posterior being evaluated. Because Mata functions can return values in this way, the Mata versions of the log-density functions are written to return the value of the log density rather than to add it to a scalar as they do in their Stata equivalents. This slightly different approach is not essential, but it makes for neater programming and means that in Mata, we do not need to set the log posterior to zero before calling logdensity().

In Mata, the information to be passed to and from a function is placed in brackets following the name of the function. Our Mata function for calculating the log posterior has three such arguments: a matrix of real numbers, X, containing the data; a row vector of real numbers, b, containing the current parameter values; and a scalar, ipar, containing the number of the parameter currently being updated. This is the order expected by the functions in the libmcmc library, so we cannot change it even though, in this example, we do not use ipar. Because the function will be compiled in advance, we include the type of each argument so that the compiler knows what type of structure to expect. The type must state whether the argument will contain reals or strings and whether it is a scalar, vector, or matrix.

The Mata version of the function `logpost()` uses matrix operators for increased speed. Entire columns are extracted from `X` using notation such as `X[,1]`, and these are combined using Mata's two types of arithmetic operations. Expressions without a colon represent the conventional matrix operations; for instance, `X*Y` is standard matrix multiplication and requires matching numbers of columns in `X` and rows in `Y`, while `3*Y` multiplies each element of `Y` by 3. On the other hand, colon operations, such as `X:*Y`, are performed element by element and require that `X` and `Y` have matching sizes, although the constraints on size are less strict than for the standard matrix operations, and in some circumstances, the colon commands will cycle over rows and columns when the sizes are mismatched. The use of these matrix operators will be more efficient than using a loop, even though loops in Mata are much faster than loops in Stata. So when we calculate the probability of success, `p`, we include terms such as `b[2]*X[,3]`, which creates a vector by multiplying every element of column 3 copied from `X` by the scalar coefficient `b[2]`. However, when we want to add the scalar `b[1]` to the vector, we use `b[1]:+b[2]*X[3]` so that `b[1]` is added to every element of the vector. See **`help m2_op_arith`** and **`help m2_op_colon`** for a fuller description of the two types of matrix operators.

Finally, the Mata version of `logdensity()` is used to calculate the components of the log posterior, and the sum is returned using the `return()` function. The arguments passed to `logdensity()` must be a string denoting the distribution, the scalar or variable that follows the distribution, and finally the parameters. As with the Stata version of this function, the value is summed over all rows whenever the quantity that follows the distribution is a vector. In this example, the log density of the binomial distribution is summed over the rows of `X[,1]` (containing `y`); the individual normal log priors, which happen to have the same form, are summed over the two elements of `b` to give the joint log prior.

In this Mata code, `p` is used to hold the calculated probabilities. We know that `p` will be a column vector with one row for each observation, but we have not told Mata that in the code, and so the compiler is left to deduce it for itself. This is an example of using the nonstrict version of Mata, and to get the code to compile, we would need to precede the Mata code by typing the Stata command

```
set matastrict off
```

In fact, Mata's compiler is very clever at deducing the forms of undefined matrices, but we could save it that task by using the strict form of Mata obtained when `matastrict` is `on`. In that case, we have to explicitly tell Mata what `p` is intended to contain by using the following code:

begin: `matastrict` log posterior

```
set matastrict on
mata:
mata clear
real scalar logpost(real matrix X,real rowvector b,real scalar ipar)
{
  real colvector p

  p = invlogit(b[1]:+b[2]*X[,3])
  return(logdensity("binomial",X[,1],X[,2],p) +logdensity("normal",b,0,4))
}
end
```

end: `matastrict` log posterior

Using the strict form of coding has two potential advantages: first, it enables the compiler to pick up errors in the code such as the creation of a square matrix when a column vector was intended; and second, it enables the compiler to create slightly faster code because it does not need to deduce the forms of the derived matrices and thus can compile using the most appropriate algorithms. In practice, the difference in speed between the strict and nonstrict versions of Mata is tiny, so the main advantage of the strict version is the ability of the compiler to detect programming errors. The strict form is clearly best practice and is especially useful when writing complex programs.

Obviously, there is much more to Mata. The manual is well over 600 pages, but this example is a template for most MCMC analyses, and further refinements will be added to the code only when needed.

7.3 Case study 6: Revisited

In chapter 6, we saw how one can validate a piece of Bayesian code by analyzing repeated samples of data simulated from parameters drawn at random from the priors. The limitation of this method is the time that it takes to run. Using Stata/SE, the ordinal regression program written in Stata code took almost exactly 45 minutes to analyze 20 randomly generated datasets. This is just the type of analysis that benefits from being run in Mata. The code below runs exactly the same analysis but evaluates the log posterior using a Mata function and avoids the use of `logdensity()`. It takes about 50 seconds to complete the same analysis, which is quicker than Stata by a factor of just over 50 times. To save space, we omit a block of Stata commands already given in section 6.3.

begin: code validation

```
mata:
mata clear
mata set matastrict off
real scalar logpost(real matrix X,real rowvector b,real scalar ipar)
{
    alpha = J(4,1,0)
    alpha[1] = b[1]
    alpha[2] = alpha[1] + b[4]
    alpha[3] = alpha[2] + b[5]
    alpha[4] = alpha[3] + b[6]
    beta  = J(3,1,0)
    beta[2]  = b[2]
    beta[3]  = b[3]
    p = J(15,1,0)
    cp = J(4,1,0)
    for (k=1;k<=3;k++) {
       cp = invlogit(alpha:+beta[k])
       p[5*k-4] = cp[1]
       p[5*k-3] = cp[2] - cp[1]
       p[5*k-2] = cp[3] - cp[2]
       p[5*k-1] = cp[4] - cp[3]
       p[5*k]   = 1 - cp[4]
    }
    return(sum(X[,1]:*log(p)) - 0.5*b[1]*b[1] - 2*b[2]*b[2] - 2*b[3]*b[3]
        + 4*log(b[4])-5*b[4] + 4*log(b[5])-5*b[5] + 4*log(b[6])-5*b[6])
}
end


tempname rf
postfile `rf´ alpha1 q1 m1 sd1 se1 beta2  q2 m2 sd2 se2 ///
              beta3  q3 m3 sd3 se3 delta2 q4 m4 sd4 se4 ///
              delta3 q5 m5 sd5 se5 delta4 q6 m6 sd6 se6 ///
              using ordinal_valid.dta,replace
set seed 639125
forvalues rep=1/20 {
   quietly drop _all
   quietly set obs 174
   generate trt = 1 + (_n>66) + (_n>119)
   generate u = runiform()
...
   matrix X = J(15,1,0)
   forvalues i=1/3 {
      forvalues j=1/5 {
         quietly count if trt == `i´ & appetite == `j´
         matrix X[5*`i´- 5  + `j´,1] = r(N)
      }
   }
   matrix list X
   matrix b = (`alpha1´,`beta2´,`beta3´,`delta2´,`delta3´,`delta4´)
   matrix s = J(1,6,0.5)
   mcmcrun logpost X b using temp.csv,              ///
      samp(3(mhsnorm, sd(s)) 3(mhslogn, sd(s)))     ///
      burnin(1000) adapt updates(20000) thin(4)     ///
      replace par( alpha1 beta2 beta3 delta2-delta4) ///
      dots(0) mata data(X s b) jpost
   import delimited temp.csv, clear
   local n = _N
```

```
        local postit ""
        foreach v of varlist a* b* d* {
           quietly count if `v´ < ``v´´
           local q`v´ = (r(N)+0.5)/(`n´+1)
           quietly mcmcstats `v´
           local postit "`postit´ (``v´´) (`q`v´´) (`r(mn1)´) (`r(sd1)´) (`r(se1)´)"
        }
        post `rf´ `postit´
    }
    postclose `rf´

    use ordinal_valid.dta, clear
    local i = 0
    foreach v of varlist q* {
        local ++i
        generate z`i´ = invnormal(q`i´)
        generate c2`i´ = z`i´*z`i´
        quietly summarize c2`i´
        display %6s "`v´" %8.3f `r(sum)´ %8.4f chi2tail(_N,`r(sum)´)
    }
```

end: code validation

The structure of the code for the function `logpost()` is essentially the same as the Stata version except that the 3×5 table of data is stored in a 15×1 column vector called `X`. The only Mata syntax that we have not used previously involves the `J(n,m,v)` function for defining an `n` $\times$ `m` matrix with all elements equal to `v` and the `for` loop, in this case, a loop in which `k` runs from 1 to 3 in steps of one.

7.4 Case study 7: Germination of broomrape

Crowder (1978) described the analysis of an experiment on the germination of broomrape. In a 2×2 unbalanced factorial experiment, 21 plates were covered in one of two types of root extract, bean, or cucumber. Then seeds from one of two varieties of the Egyptian broomrape were placed on the plates and monitored to see whether they germinated. Broomrape is a parasitic plant and germinates in the presence of the roots of plants that it can parasitize. For each plate, we have the number of seeds (`n`) and the number that germinated (`y`), and these data are shown in table 7.1.

Table 7.1. Proportions of seeds germinating on each of the 21 plates

O. aegyptiaco 75				O. aegyptiaco 73			
bean		cucumber		bean		cucumber	
10/39	23/62	5/6	53/74	8/16	10/30	3/12	22/41
23/81	26/51	55/72	32/51	8/28	23/45	15/30	32/51
17/39		46/79	10/13	0/4		3/7	

These data can be stored in Stata variables: `y`, `n`, `seed` (0 = variety 75, 1 = variety 73), and `root` (0 = bean, 1 = cucumber). A natural model for them would be a logistic regression. However, there is a strong possibility of a plate effect that would cause overdispersion in a standard logistic regression, and one way to allow for this is to include a normally distributed random effect with a different value for each plate. If we include the interaction between the root extract and the variety, the model becomes

$$\begin{aligned} y_i &\sim B(p_i, n_i) \qquad i = 1, \ldots, 21 \\ \text{logit}(p_i) &= \alpha_0 + \alpha_1 \text{seed} + \alpha_2 \text{root} + \alpha_{12} \text{seed} * \text{root} + u_i \\ u_i &\sim N(0, \sigma) \end{aligned}$$

It is worth noting that the random effect in this model has to be interpreted with some care. There are no repeat observations with the same value of the random effect, so the size of u_i is in part a measure of differences between plates and in part a reflection of the appropriateness or otherwise of our choice of link function.

Setting informative priors for this experiment requires a discussion with an expert on broomrape germination. For instance, it is important to know whether the root extracts were selected because they represent suitable hosts for these varieties or because the experimenters wanted to find out whether they were suitable hosts. In the latter case, a zero germination rate would be a realistic possibility, while in the former, it would not. Assuming that some germination is expected from both root extracts, then linear predictors ranging between ± 3 lead to germination rates between 5% and 95%. So we might safely assume that the coefficients in the linear predictor are all well within the range ± 4 and that the plate effects, u_i, are probably not bigger than ± 2, which suggests a standard deviation not much greater than 1. Possible choices are

$$\begin{aligned} \alpha_0, \alpha_1, \alpha_2, \alpha_{12} &\sim N(0, 2) \\ \tau &\sim G(2, 2) \\ \sigma &= 1/\sqrt{\tau} \end{aligned}$$

The prior on the precision, τ, gives reasonable levels of support to values in the range (0.3,13), so by implication, there is support for standard deviations, σ, between about 0.28 and 1.8. Importantly, this prior forces the model away from a solution in which the random effect is negligible. Many statisticians are nervous of priors of this type and would opt for a much wider distribution that allowed the standard deviation to approach closer to zero. In this example, such a wide prior results in a wide posterior because the size of the random effect is so poorly defined by the data.

This model could be fit using the Metropolis–Hastings (MH) algorithm with a normal random-walk proposal with centered covariates using the following commands:

begin: broomrape analysis

```
use broomrape.dta, clear
summarize seed
local m1 = r(mean)
generate x1 = seed - `m1´
summarize root
local m2 = r(mean)
generate x2 = root - `m2´
generate x12 = root*seed
summarize x12
local m12 = r(mean)
replace x12 = x12 - `m12´
set matastrict off
mata:
mata clear
real scalar logpost(real matrix X,real rowvector b,real scalar ipar)
{
  sd = 1/sqrt(b[5])
  eta = b[1]:+b[2]*X[,3]+b[3]*X[,4]+b[4]*X[,5]+b[|6\26|]´
  p = invlogit(eta)
  return(logdensity("binomial",X[,1],X[,2],p)
       + logdensity("normal",b[|6\26|]´,0,sd)
       + logdensity("normal",b[|1\4|],0,2) + logdensity("gamma",b[5],2,2))
}
end

matrix theta = J(1,26,0)
matrix theta[1,5] = 1
mcmcrun logpost X theta using temp.csv, samp(4(mhsnorm, sd(0.2))  ///
   (mhslogn, sd(0.2)) 21(mhsnorm, sd(0.2))) replace burn(1000)     ///
   update(10000) jpost  dots(0) par(a0 a1 a2 a12 tau u1-u21)       ///
   mata data(X=(y n x1 x2 x12) theta)
import delimited temp.csv, clear
replace a0 = a0 - `m1´*a1 - `m2´*a2 - `m12´*a12
generate sd = 1/sqrt(tau)
mcmcstats a* sd
```

end: broomrape analysis

The code follows much the same pattern as the template developed in section 7.2. The covariates have been centered within Stata before being passed to Mata to improve mixing. When preparing the code, one should note the order of the parameters within theta. Here we have opted for a0, a1, a2, a3, s, u1, u2, ..., u21. This means that 0 is a possible starting value for all the parameters except number five, which has its initial value set to 1.

Two new Mata commands are used in this code: a single quote mark (´), which is used to denote the transpose of a matrix, and y[|a\b|], which is used to denote a vector formed by extracting elements a to b from the vector y.

The Mata code uses MH sampling to create 11,000 simulations, of which the first 1,000 are discarded as the burn-in. However, as figure 7.1 shows, mixing is not particularly good.

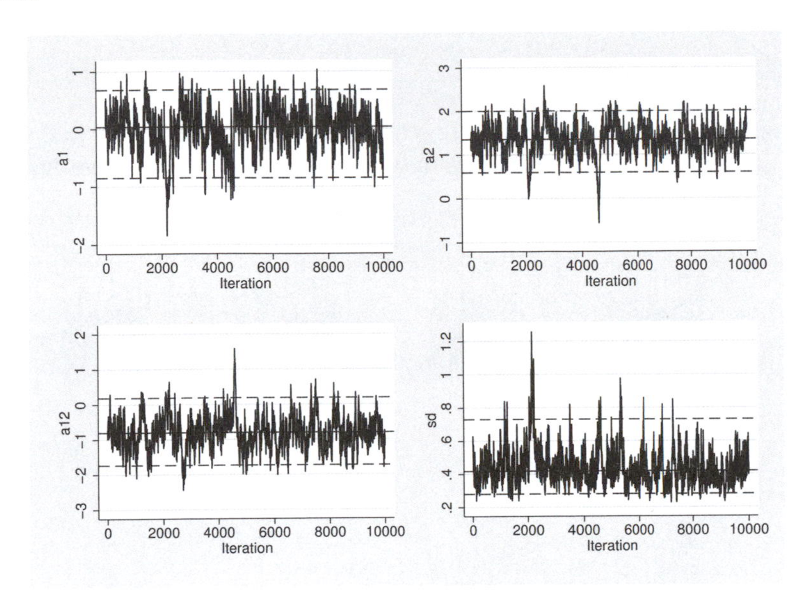

Figure 7.1. Trace plots for the initial broomrape analysis

Before we attempt to summarize the results, let's consider:

- Would convergence be improved by better proposal distributions?
- Can the code be made more efficient?
- Would convergence be improved by a different parameterization?
- Would other samplers perform better?

7.4.1 Tuning the proposal distributions

As we saw in chapter 3, the standard deviations of the proposal distributions can be improved during the burn-in to achieve an acceptance rate close to 50% by simply adding the `adapt` option to `mcmcrun`. This works equally well with Mata functions.

begin: proposal adjustment

```
matrix s = J(1,26,0.2)
mcmcrun logpost X theta using temp.csv, samp(4(mhsnorm, sd(s))  ///
   (mhslogn, sd(s)) 21(mhsnorm, sd(s))) replace burn(1000)      ///
   update(10000) jpost  dots(0) par(a0 a1 a2 a12 tau u1-u21)    ///
   adapt mata data(X=(y n x1 x2 x12) theta s)
```

end: proposal adjustment

In this code, the standard deviations are stored in the row vector `s` and all start at 0.2. The resulting improvement in mixing is quite marked, as can be seen in figure 7.2.

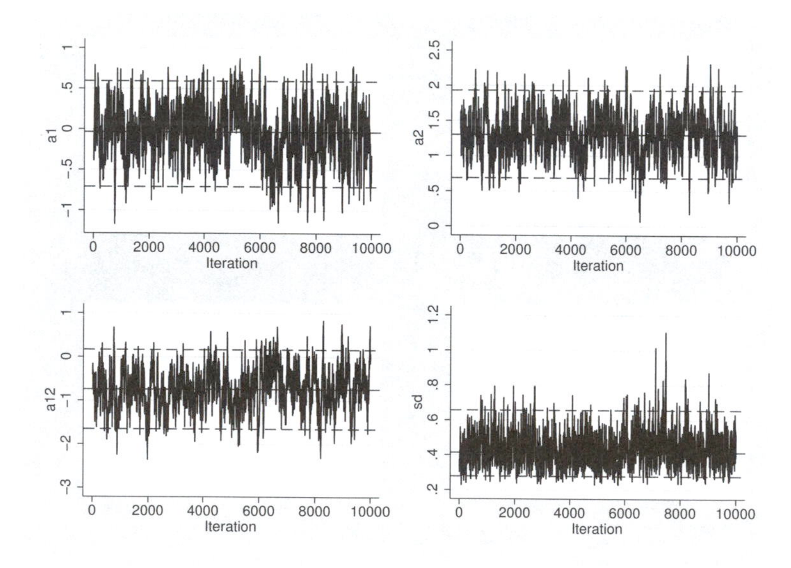

Figure 7.2. Trace plots for the broomrape analysis after tuning of the proposal standard deviations

7.4.2 Using conditional distributions

Replacing the joint posterior by the appropriate conditional distributions will usually reduce calculation time because many of the terms in the joint posterior will be constant within a particular conditional distribution and so can be omitted. However, because the form of the conditional distribution changes according to the parameter being updated, one cannot use the value of `logp` from the previous step as the current value for the next parameter. Consequently, each MH step will require two function evaluations instead of one. There is a tradeoff, but in large problems, the gain in execution time from using the conditional distributions is usually quite marked. To switch to conditional distributions, we must make two changes: first, to the function `logpost()`, and second, to `mcmcrun`, which must be run without the `jpost` option. The changes to `logpost()` are shown below.

begin: conditionals

```
mata:
mata clear
real scalar logpost(real matrix X,real rowvector b,real scalar ipar)
{
  sd = 1/sqrt(b[5])
  if(ipar < 5) {
     mu = b[1]:+b[2]*X[,3]+b[3]*X[,4]+b[4]*X[,5]+b[|6\26|]'
     p = invlogit(mu)
     return(logdensity("binomial",X[,1],X[,2],p)
          + logdensity("normal",b[ipar],0,4))
  }
  else if(ipar == 5) {
     return(logdensity("normal",b[|6\26|],0,sd)
          + logdensity("gamma",b[5],2,2))
  }
  else {
     i=ipar-5
     mui = b[1]+b[2]*X[i,3]+b[3]*X[i,4]+b[4]*X[i,5]+b[ipar]
     pi = invlogit(mui)
     return(logdensity("binomial",X[i,1],X[i,2],pi)
          + logdensity("normal",b[ipar],0,sd))
  }
}
end
```

end: conditionals

7.4.3 More efficient computation

Another way to improve the speed of the algorithm is to abandon `logdensity()` and adopt more efficient computation. In the code used thus far, we have calculated the linear predictor, converted it into a probability, and then used that probability to calculate the log of the binomial likelihood. Simple algebra shows that this process can be sped up by going directly from the linear predictor to the binomial distribution. If the linear predictor is η, then

$$p = \frac{e^{\eta}}{1+e^{\eta}} \quad 1-p = \frac{1}{1+e^{\eta}}$$

so that we can rewrite the log-binomial density without its unnecessary constants as follows:

$$y\eta - n\log(1+e^{\eta})$$

The code below shows a more efficient program for the conditional posterior distribution.

begin: more efficient logdensities

```
mata:
mata clear
real scalar logpost(real matrix X,real rowvector b,real scalar ipar)
{
  sd = 1/sqrt(b[5])
  if(ipar < 5) {
     mu = b[1]:+b[2]*X[,3]+b[3]*X[,4]+b[4]*X[,5]+b[|6\26|]'
     p = invlogit(mu)
     return(sum(X[,1]:*mu:-X[,2]:*log(1:+exp(mu))) - b[ipar]*b[ipar]/32)
  }
  else if(ipar == 5) {
     return(-0.5*b[5]*b[|6\26|]*b[|6\26|]' -21*log(sd) + log(b[5]) -b[5]/2)
  }
  else {
     i=ipar-5
     mui = b[1]+b[2]*X[i,3]+b[3]*X[i,4]+b[4]*X[i,5]+b[ipar]
     return(X[i,1]*mui-X[i,2]*log(1+exp(mui)) -0.5*b[5]*b[ipar]*b[ipar])
  }
}
end
```

end: more efficient logdensities

7.4.4 Hierarchical centering

Centering covariates is a routine method of reparameterization that almost always improves convergence. We have already reduced the correlation between the intercept, a_0, and the other regression coefficients by subtracting the covariate means. Although this is always sensible, it does not help with the correlations that affect the random effects. Further improvement in mixing can sometimes be achieved by using a reparameterization known as hierarchical centering, which involves writing the model so that the random effects have a parameterized mean rather than a zero mean. In the logistic regression example, the original model was

$$\begin{aligned} r_i &\sim \text{bin}(p_i, n_i) \qquad i = 1, \ldots, 21 \\ \text{logit}(p_i) &= \alpha_0 + \alpha_1 x1 + \alpha_2 x2 + \alpha_{12} x12 + u_i \\ u_i &\sim N(0, \sigma) \end{aligned}$$

but an equivalent model with the new parameterization would be

$$\begin{aligned} r_i &\sim \text{bin}(p_i, n_i) \qquad i = 1, \ldots, 21 \\ \mu &= \alpha_0 + \alpha_1 x1 + \alpha_2 x2 + \alpha_{12} x12 \\ \text{logit}(p_i) &= v_i \\ v_i &\sim N(\mu, \sigma) \end{aligned}$$

In the second formulation, the new parameters v_i should mix more freely than did the u_i, but if needed, the original random effects u_i can still be deduced from the output by subtracting the corresponding estimates of μ. The change to the code for `logpost` is straightforward and is shown below.

begin: hierarchical centering

```
mata:
mata clear
real scalar logpost(real matrix X,real rowvector b,real scalar ipar)
{
  sd = 1/sqrt(b[5])
  if(ipar < 5) {
     mu = b[1]:+b[2]*X[,3]+b[3]*X[,4]+b[4]*X[,5]
     v = b[|6\26|]´
     return(-0.5*b[5]*sum((v:-mu):*(v:-mu)) - b[ipar]*b[ipar]/32)
  }
  else if(ipar == 5) {
     mu = b[1]:+b[2]*X[,3]+b[3]*X[,4]+b[4]*X[,5]
     v = b[|6\26|]´
     return(-0.5*b[5]*sum((v:-mu):*(v:-mu)) - 21*log(sd) + log(b[5]) - b[5]/2)
  }
  else {
     i=ipar-5
     mui = b[1]+b[2]*X[i,3]+b[3]*X[i,4]+b[4]*X[i,5]
     return(X[i,1]*b[ipar]-X[i,2]*log(1+exp(b[ipar]))
         - 0.5*b[5]*(b[ipar]-mui)*(b[ipar]-mui))
  }
}
end
```

end: hierarchical centering

7.4.5 Gibbs sampling

When a gamma prior is used for the precision, τ, of the normally distributed random effects, its conditional posterior will also follow a gamma distribution, and so it is possible to create an exact Gibbs sampler for τ. This not only will be quicker than using MH but also should improve the mixing. As we saw in section 4.3, extracting the terms in τ from the joint posterior in the hierarchically centered formulation and ignoring constants shows that τ follows a gamma distribution:

$$G\left[12.5, 2\left\{1+\sum_{i=1}^{21}(v_i-\mu_i)^2\right\}^{-1}\right]$$

To implement this within a call to `mcmcrun` requires one extra Mata function and a change to the list of samplers.

```
mata:
mata clear
void myGibbs(real scalar logp,pointer(function) userprog,
         real matrix X, real rowvector b,real scalar ipar)
{
   mu = b[1]:+b[2]:*X[,3]:+b[3]:*X[,4]:+b[4]:*X[,5]
   r = b[|6\26|]´-mu
   b[5] = rgamma(1,1,12.5,2/(1+r´*r))
}
end
matrix theta = J(1,26,0)
matrix theta[1,5] = 1
matrix s = J(1,26,0.2)
mcmcrun logpost X theta using temp/temp.csv, samp(4(mhsnorm, sd(s)) ///
   (myGibbs) 21(mhsnorm, sd(s))) replace burn(1000)                ///
   update(10000) dots(0) par(a0 a1 a2 a12 tau u1-u21) adapt         ///
   mata data(X=(y n x1 x2 x12) theta s)
```

Because there is no standard sampler called `mygibbs()`, `mcmcrun` assumes that the user has defined one and calls it with arguments `X` and `theta`.

7.4.6 Slice, Griddy, and ARMS sampling

As discussed in chapter 4, black-box samplers such as the slice sampler, griddy sampler, and the adaptive rejection Metropolis sampling (ARMS) sampler require more function evaluations at each step than does the MH sampler and so will be slower to complete the same number of simulations. However, they all sample from the full conditional distribution and should mix better; thus they may require a shorter run length, which sometimes makes them quicker than MH.

7.4.7 Timings

The time needed to analyze the broomrape data will obviously depend on the computer used, but relative times between the different algorithms seem fairly stable across computers. The table below gives the timings for various samplers based on a run length of 50,000. In each case, the run was preceded by a burn-in of length 1,000, during which the proposal standard deviations of the MH samplers were adapted. As a rough guide, using Stata rather than Mata with equivalent coding of the log posterior takes at least 10 times as long.

Because some samplers mix better than others, the time taken to complete a run of a fixed length is not the best guide to performance; a poorly mixing sampler will eventually have to run longer. To give a fairer comparison, the right-hand column of table 7.2 gives an estimate of how long it would take to create a run for which the sems for all the regression coefficients would be less than 0.01.

Table 7.2. Times in seconds to run different analyses with a run length of 50,000; the second time in each pair is an estimate of how long it would take to run a chain that has standard errors below 0.01 for all regression coefficients.

Run	Sampler	Log-posterior coding	$n = 50{,}000$	sem = 0.01
1	MH	Joint using logdensity	40.1s	85.5s
2	MH	Conditional using logdensity	36.0s	82.1s
3	MH	Conditional without logdensity	13.9s	30.0s
4	MH	As run 3 plus hierarchical centering	12.5s	19.8s
5	MH	As run 4 plus Gibbs for tau	12.0s	17.3s
6	Griddy	As run 5; Griddy n=5 metropolis	67.6s	63.6s
7	ARMS	As run 5; ARMS $n = 5$	197.0s	142.3s
8	Slice	As run 5	34.3s	18.8s

With the runs that used MH samplers, the efficiency of the coding of `logpost()` is of critical importance because the other parts of the MCMC calculation are relatively quick. Timings obtained from the tuned MH analyses using the conditional posteriors (run 2) suggest that over 75% of the run time is spent calculating `logpost()` and that the remainder is spent on tasks such as writing to the file, generating new proposals, and testing their acceptance. With more complex problems, the time needed to calculate `logpost()` is likely to increase relative to the time needed for the other tasks, which makes efficient programming even more important. The timings in table 7.2 show that the biggest jump in performance comes when the calls to `logdensity()` are replaced by more efficiently written code that does not calculate unnecessary constants in the densities. Hierarchical centering does not make much difference to the time taken to run 50,000 iterations, but it does improve the mixing, so the sem is reduced, and a shorter run will be required to give any specified accuracy.

Slice sampling is very competitive with the MH algorithm; it takes slightly longer to run but has much better mixing, and so the time needed to reduce the sem of the regression coefficients to 0.01 is about the same. Griddy sampling with $n = 5$ and a Metropolis step to ensure that it converges to the correct posterior is about three times slower than slice sampling or MH. ARMS with $n = 5$ is even slower and does not seem a practical algorithm for routine use. This is not to say that griddy sampling and ARMS will never be useful; in this example, the initial grid was placed without any special knowledge of the likely position of the posterior and therefore needed to cover a wide range of values. There may be circumstances in which the initial grid could be placed more accurately; in such cases, these methods might prove useful.

The larger the problem, the more critical the timing becomes. A saving of 20% may not be worth the programming effort for a problem that completes in 60 seconds but could be of real importance if each run takes 5 hours. The messages from these analyses are the following:

- Mata is at least 10 to 20 times faster than Stata.
- Mixing must be considered alongside the speed per iteration.
- Most of the computation time goes in calculating the log posterior.
- Conditional posteriors will usually be faster than the joint posterior.
- Efficient calculation of the log posterior can halve the computation time.
- Recognized conditional distributions should be programmed with exact Gibbs samplers.
- MH and slice sampling are the fastest algorithms for nonstandard distributions.

7.4.8 Adding new densities to logdensity()

The code for evaluating the log posterior uses prepared functions for calculating the log density of the various distributions. These are part of the `logdensity()` function stored in the Mata library, `libmcmc`. Should you want to add a new density to the library, you can edit the code that generates the library and then recompile it. The following example shows the section of code used for calculating the log of the gamma density:

begin: `logdensity()`

```
mata:
real scalar logdensity(
    string scalar distribution,
    real matrix y,
    real matrix p1,
  | real matrix p2,
    real matrix p3,
    real matrix p4)
{
    if(distribution == "bernoulli") {
        ...
    }
        ...
    else if(distribution == "gamma") {
        if(args() < 4 || args() > 5) exit(3000)
        if(args() == 5) w = p3; else w = 1
        return(sum(w:*((p1:-1):*log(y):-y:/p2:-p1:*log(p2):-lngamma(p1))))
    }
        ...
}
mata mlib add libwb logdensity()
end
```

end: `logdensity()`

The code in `logdensity()` uses elementwise matrix operators, which allow it to work with any mixture of matrix or scalar parameters provided that all matrices are of the same size. So unlike the Stata equivalent, there is no need to have separate code for scalars and vectors. The parameters listed after the | sign are optional, which allows the numbers of parameters to vary depending on the distribution. If there are more specified arguments than there are parameters in the distribution, then the next argument in the list is treated as a weight; this could be used to exclude some observations by setting their weight to zero. The returned value is the weighted sum of the log posterior over all observations.

7.5 Further reading

Full details on using Mata are given in the Stata manuals, but much of that material is explained more fully in Baum (2009). Several important topics have been covered in *Stata Journal* articles called *Mata matters*; see, for instance, Gould (2006, 2007, 2008). A lot of information is also available on the Stata website, http:///www.stata.com.

7.6 Exercises

1. The data in the table below were originally used to illustrate an analysis in Beitler and Landis (1985). They show the results of a randomized clinical trial conducted in eight clinics. Patients with nonspecific infections were randomized to receive a topical cream (1) or to be given no treatment (0), and the proportion cured was then noted.

Clinic	Treatment	Cured	Treated
1	1	11	36
1	0	10	37
2	1	16	20
2	0	22	32
3	1	14	19
3	0	7	19
4	1	2	16
4	0	1	17
5	1	6	17
5	0	0	12
6	1	1	11
6	0	0	10
7	1	1	5
7	0	1	9
8	1	4	6
8	0	6	7

Because the patient mix attending the different clinics varies, one might anticipate that the cure rate, even in the absence of treatment, also varies. For this reason, one should compare treatments within clinics. Assuming that the size of the benefit is common across clinics, we might use a model such as

$$\begin{aligned} y_{ij} &\sim \text{bin}(p_{ij}, n_{ij}) \\ \text{logit}(p_{ij}) &= \alpha_i + \beta x_{ij} \\ \alpha_i &\sim N(\mu, \sigma) \end{aligned}$$

where y_{ij} represents the number cured in clinic i with treatment j, x_{ij} equals 0 or 1 and represents the treatment, and n_{ij} the number treated. The parameters, α_i, represent the baseline cure rates, and β measures the common effect of treatment. A hierarchical distribution is placed over α_i to improve the estimation in the clinics that recruited few patients.

a. Decide whether to work in terms of the standard deviation, σ, or its equivalent precision. Specify your own priors. Then write a Mata program to calculate the log of the product of joint likelihood times prior for this model using calls to `logdensity()`, and use it to fit the model with your own choice of samplers. Check convergence and summarize the treatment effect.
b. Rewrite the Mata program with efficiently calculated conditional distributions that do not rely on `logdensity()`, and compare timings with the previous version.
c. An alternative model for these data allows the treatment effect to vary across clinics, perhaps placing a hierarchical normal distribution over the β_i with mean m and standard deviation s. Specify realistic priors for m and s, and then modify your program for the conditional analysis so that it fits this new model. Compare the results with those obtained with a very wide prior on s. Is there any evidence that the treatment effect does indeed vary across clinics?

2. Case study 8 (section 8.4) presents data on the lengths of sea cows of various ages. In that chapter, these data are analyzed using nonlinear model fit by WinBUGS. Analyze the data using the same model by deriving your own Gibbs samplers for the standard conditional distributions of α, β, and τ together with an MH sampler for γ. Program the analysis in Mata. (Note that the model given in section 8.4 uses WinBUGS parameterization for the gamma and normal distributions. See the appendix for details.)

8 Using WinBUGS for model fitting

8.1 Introduction

Gibbs sampling and the Metropolis–Hastings algorithm are convenient ways of fitting Bayesian models provided that the algorithms can be made efficient. This requires fast evaluation of the conditional posterior distributions and careful tuning of the algorithms to produce a well-mixing chain. Many of these efficiency problems are solved automatically by using WinBUGS, a free program written for the Windows operating system that incorporates a range of samplers and that has the intelligence to select the best and to tune it for any given problem. As a result, using WinBUGS is often the simplest and most efficient way of fitting a Bayesian model.

The chief disadvantage of WinBUGS is that it does not have all the tools for data handling and graphics that Stata users take for granted. Consequently, a good way of working is to hold the data in Stata, to export the model-fitting problem to WinBUGS, and to read the resulting Markov chain Monte Carlo (MCMC) simulations back into Stata for further processing. Several downloadable ado-files with names beginning `wbs` control communication between Stata and WinBUGS.

8.2 Installing the software

If you are new to WinBUGS, then you should probably use the OpenBUGS version of the program. WinBUGS started as a program called Bayesian inference using Gibbs sampling (BUGS), which was developed at the MRC Biostatistics Unit in Cambridge, United Kingdom. Later the same group produced a more interactive version of the program called WinBUGS. This Windows-based program went through several upgrades but is now frozen at version 1.4.3. Future development has moved to the open-source, OpenBUGS project. WinBUGS and OpenBUGS have an almost identical user interface and have been tested to ensure that they give the same results for a wide range of models, but the OpenBUGS version does have a few useful extra features. It also has the big advantage that it offers a version of the program that runs under Linux. However, to simplify the following description, we will concentrate on the Windows versions and only afterward describe the changes required when running OpenBUGS under Linux.

8.2.1 Installing OpenBUGS

OpenBUGS can be downloaded from its website, http://www.openbugs.net/. This website also includes links to documentation, to different projects linked to OpenBUGS, and to a page describing the small differences between OpenBUGS and WinBUGS. OpenBUGS is an open-source program distributed under a GNU general public license, so after reading the terms and conditions of the license, you do not need to do anything else to run the program.

Once the program is installed, you need to tell Stata where the executable is on your computer. For users who do not store OpenBUGS on the Windows search path, the `wbsrun` command has an option that enables the user to specify the location of the executable; however, when many analyses will be run, it is more convenient to store the path in a text file and to allow the calling program to look it up as required. To do this, create a text file called `executables.txt` in your `PERSONAL` folder. The location of the `PERSONAL` folder can be obtained by typing the command `sysdir` into Stata. The executables file may include information about other programs, including the path to the frozen version of WinBUGS, but must include the line

```
OPENBUGS,"path/openbugs.exe"
```

where *path* is replaced by the path to the OpenBUGS executable on your computer. If the path contains spaces, then the quotes around the path are essential.

8.2.2 Installing WinBUGS

The frozen version of WinBUGS can be downloaded from its homesite at the MRC Biostatistics Unit in Cambridge, United Kingdom, http://www.mrc-bsu.cam.ac.uk/bugs/. This website includes full instructions for installing the software, obtaining the free license key, and installing patches. It also gives a lot of other information about WinBUGS and associated projects, some of which will be described later in this chapter.

WinBUGS is downloadable as version 1.4, but a patch is available to upgrade the software to version 1.4.3. It is important to install this patch because it includes some critical bug fixes. Once version 1.4 has been downloaded and unzipped, run WinBUGS interactively, and open the patch file using WinBUGS's `FILE` menu. The patch file contains instructions for its own installation, which merely involves selecting an option from WinBUGS's `TOOLS` menu. WinBUGS also requires the installation of a license key using a process that follows much the same pattern; the key is obtained free from the website as a downloadable text file. It is opened in WinBUGS and contains its own instructions for decoding using the `TOOLS` menu. WinBUGS licenses were previously issued annually, but version 1.4.3 is frozen with future development transferred to OpenBUGS, so the current license is perpetual.

If WinBUGS is not installed on the Windows search path, you will need to tell Stata where the WinBUGS executable is on your computer. As for OpenBUGS, a convenient way of telling the `wbs` commands where to find WinBUGS is to include a line in the `executables.txt` file stored in your `PERSONAL` folder. In this case, it takes the form

```
WINBUGS,"path/winbugs14.exe"
```

where *path* is replaced by the path to WinBUGS on your computer.

If you wish, you can install both OpenBUGS and WinBUGS on the same computer without causing any clashes.

8.3 Preparing a WinBUGS analysis

Even though we recommend that you use OpenBUGS, the following sections refer to the use of WinBUGS because this is the name by which the software is best known. Because the syntax of WinBUGS and OpenBUGS are so similar, most of the description is equally applicable to either program, so we will only distinguish between them on the rare occasions when they differ.

Specifying an MCMC analysis in WinBUGS requires three components: a description of the model, the data, and the initial values. When one runs WinBUGS interactively, it is usual, though not essential, to place these three components in a single compound document called an `odc` file. WinBUGS's `odc` files can contain in the one place the three components of the problem, the results, the graphs, and the descriptive text. When interactively processing one component, the relevant part of the file is highlighted before the appropriate menu item is chosen. However, running WinBUGS from within Stata requires a different approach: the three components of the problem are placed in separate text files, and a script file is prepared that uses a batch processing language to control the options that would otherwise have been selected interactively. Within the batch commands, WinBUGS can be told to write the simulated parameter estimates to another text file from which they can be read into Stata.

The following sections describe the structure of the four text files for the model, data, initial values, and script and explain how they can be prepared and called from within Stata. Using this approach, WinBUGS is just a black-box for computation while the user works entirely within Stata.

8.3.1 The model file

The model must be described in a text file using the WinBUGS syntax. This file could be prepared in any text editor, but for Stata users, it will probably be most convenient to prepare it in the Do-file Editor. The model file contains instructions for calculating the value of the posterior distribution; in many ways, it is not unlike the `logpost` program used in earlier chapters, except that WinBUGS does all the work of turning the code into an efficient algorithm, so users do not need to worry so much about the efficiency of their specification of the problem.

The WinBUGS language is relatively straightforward and is described in a manual that is accessible when WinBUGS is run interactively. Perhaps more helpful than the manual is a set of accompanying examples that show the WinBUGS code for tackling over 30 sample problems. Most people learn to use WinBUGS by copying these examples.

In the WinBUGS syntax, all the standard distributions are identified by a keyword that begins with the letter "d"; thus `dnorm()` refers to a normal distribution. These keywords are listed in the WinBUGS manual but are also given alongside the formulas for the distributions in the appendix to this book. Thus the model description might include

```
y ~ dnorm(0,1)
```

to show that `y` follows a standard normal distribution with mean 0 and precision 1. Care must be taken with the parameterization of the distributions because sometimes, WinBUGS uses a different parameterization than Stata. The normal distribution, for example, is specified in terms of the precision (1/variance), while Stata uses the standard deviation.

Loops are created using the construct

```
for (i in 1:10) {
...
}
```

and standard mathematical functions such as `log()` and `exp()` are denoted in the usual way. One slight oddity of WinBUGS is to allow `log()` and `logit()` to appear on the left-hand side of an assignment. There is no inverse logit function, so one needs to write

```
logit(y) <- a + b*x
```

where in Stata, we would write `y=invlogit(a+b*x)`. Notice that WinBUGS uses `<-` for assignment in place of the more usual equals sign. To be exact, OpenBUGS version 3.2.2 has introduced an `ilogit()` function, although placing `logit` on the left of the assignment still works and is common practice.

This small amount of syntax is enough to enable a simple model to be described. So, as an example, we will consider the logistic regression model with a random effect that was used in case study 7 to analyze the broomrape germination data. This model could be specified in WinBUGS as follows:

begin: simple WinBUGS model

```
model {
   for (i in 1:21) {
      logit(p[i]) <- a0 + a1*x1[i] + a2*x2[1] + a3*x12[i] + u[i]
      y[i] ~ dbin(p[i],n[i])
      u[i] ~ dnorm(0,tau)
   }
   a0 ~ dnorm(0,0.25)
   a1 ~ dnorm(0,0.25)
   a2 ~ dnorm(0,0.25)
   a3 ~ dnorm(0,0.25)
   tau ~ dgamma(2,0.5)
}
```

end: simple WinBUGS model

Because there are 21 observations, we need to loop over them and specify the distribution of each one using square brackets, `[]`, for subscripting. In this model description, `dbin()` refers to the binomial distribution, and `dgamma()` refers to the gamma distribution. Because WinBUGS parameterizes the normal distribution by its mean and precision, 0.25 in the priors on the regression coefficients corresponds to a standard deviation of 2, and the gamma distribution is parameterized in terms of shape and rate rather than shape and scale. Thus WinBUGS's `tau ~ dgamma(2,0.5)` corresponds to Stata's `gammaden(2,2,0,tau)` and to `logdensity` with both parameters equal to 2.

Of course, there are many more components of the WinBUGS language, but we postpone their consideration until they are needed because this example acts as a template for most simple models.

8.3.2 The data file

WinBUGS is particular about what appears in the data file and will give an error message if variables are included that are not eventually referred to in the model. This is rather inconvenient because it means that it is not possible to supply a standard data file and then only refer to the variables that are relevant for a particular analysis. Instead, a new data file must be created for each model.

Data can be written to the data file in one of two formats: either as a rectangular `array` with rows and columns of the type seen in Stata's Data Editor or as a `list` with an R-like structure. For instance, a vector of length three might be specified as a list using

```
list(y = c(10, 23, 4))
```

where `c()` denotes that the values are to be combined into a vector. Notice that the equals sign is used for assignment in the data file. WinBUGS allows multiple data files, so users can have a mixture of arrays and lists by placing them in different files.

Preparing the data for WinBUGS is one of the most awkward aspects of the package, whether it is used via Stata or interactively. To speed up this process, two downloadable programs write Stata data in the required format without users needing to concern themselves with the structure of the resulting text files. `wbslist` writes data as a list, and `wbsarray` creates a rectangular table of data. When `wbslist` or `wbsarray` is used, there is no need to understand the exact way that WinBUGS formats its data, but of course, if needed, the formats are all described in the WinBUGS manual.

To create a rectangular dataset in the file `data.txt`, simply list the Stata variables that contain the data after `wbsarray`, for instance,

```
wbsarray x1 x2 x3 using data.txt, replace
```

`wbslist` writes data taken from Stata matrices, scalars, or variables provided that the different types of data are distinguished by being placed in brackets within which the first word denotes the type of data. For instance,

```
wbslist (variable x y z) (matrix R) using data.txt, replace
```

writes the three Stata variables `x`, `y`, and `z` and the Stata matrix `R` in list format to the file `data.txt`, replacing the file if it already exists.

`wbslist` can also combine data from several Stata variables into a WinBUGS matrix by using the keyword `structure`. For instance,

```
wbslist (structure x y z, name(X)) using data.txt, replace
```

writes the three Stata variables `x`, `y`, and `z` in a form that WinBUGS reads as a matrix called `X` with three columns. The keyword `table` can be used to create structures with higher numbers of dimensions (see the help file for details). Both `wbslist` and `wbsarray` allow the user to specify the format used to write the data to the text file, and they allow `if` or `in` qualifiers so that subsets of the data may be selected.

8.3.3 The initial values file

The initial values are specified in a text file with the same list format as a data file, and they can therefore be written using the same `wbslist` command. WinBUGS will, if requested, make more than one chain of simulations for the same model, and as we saw in chapter 5, this can be useful for investigating whether chains that start from different points all converge to the same distribution. In such analyses, each chain requires its own set of initial values written to a different text file.

One aspect of `wbslist` is specifically designed for writing initial values. When the user supplies a pair of brackets without a keyword to identify a type of data, the contents of the bracket are written directly to the text file, except that anything enclosed within curly brackets, `{ }`, is evaluated. For instance,

```
wbslist (alpha=c(4{round(rnormal(),0.01)}),t={1+3}) ///
     using inits.txt, replace
```

would evaluate the Stata expression

```
round(rnormal(),0.01)
```

four times. So the resulting file `inits.txt` might read something like

```
list(alpha=c(0.23,-1.43,0.74,-0.12),t=4)
```

although the exact values in alpha would be chosen randomly and so vary from run to run. This feature enables the user to create random starting values, perhaps for running multiple chains.

Another useful feature of the syntax is the use of square brackets to select the initial values from within a Stata variable. Thus

```
wbslist (alpha=c(4{x[6]})) using inits.txt, replace
```

which fills `alpha` with 4 values taken from observations 6, 7, 8, and 9 within the Stata variable `x`.

If a file of initial values does not contain initial values for all the parameters, WinBUGS can be told to try to generate random starting values for the remainder. This often works well, but success is problem specific. For more complex models, it is usually safer to supply a complete set of initial values.

8.3.4 The script file

The script file is another text file that contains a series of one-line commands that tell WinBUGS which actions to perform, for instance, to read the model, read the data, compile the model, run a burn-in of a specified length, etc. Rather than write the file yourself, it is easier to run the `wbsscript` command that creates it for you. Thus, having written a model file, with corresponding data and initial value files all stored in the directory `c:\temp`, you can create a script file called `script.txt` by typing the following Stata command:

```
wbsscript using script.txt, model(model.txt) data(data1.txt+data2.txt) ///
      inits(init.txt) burn(1000) update(5000) thin(5) log(log.txt)     ///
      set(alpha beta) coda(mymod) path(c:\temp)
```

It is important that all of your text files referred to within the script be given a full path because you cannot guarantee that when WinBUGS runs, it will start in the same folder that you are currently using as your working directory in Stata. Files without an explicit path will therefore not be found. You can give full paths within the `model()`, `data()`, and `inits()` options, but frequently, users store these files together in the same folder. In this case, you can save a little typing by placing that path within the path option, as we have done in the example above. If the path option is omitted and a file does not have an explicit path, then it is assumed that the file will be found in the current working directory, and that path is automatically added. The output files specified by the `log()` and `coda()` options obey the same rules, so if the `path()` option is omitted and you do not give an explicit path, then they will be written to the current working directory.

The script file, `script.txt`, will be stored in Stata's current working directory unless it is explicitly given a path following the word `using`; that is, it is not affected by use of the `path()` option.

In this example, there are two data files and one set of initial values, so only one chain will be run. To run two chains, we would need to have two sets of initial values and change the `inits()` option to something like `inits(init1.txt+init2.txt)`. `wbsscript` automatically inserts a request for WinBUGS to generate random starting values for any parameters not included in the initial values file. The chain will have a discarded burn-in of 1,000 followed by 5,000 updates, but only every 5th will be stored. No matter how many parameters our model might have, only the simulations for `alpha` and `beta` will be saved, and those values will be written to text files with the prefix `mymod`.

Occasionally, you will mistype something in the model file. The result will be that when WinBUGS is called by Stata, the compilation will fail, and WinBUGS will exit quickly without creating the MCMC chain. You will see that this has happened when you look at the log file. To debug the problem, you need to identify the place where the error occurred, which means stopping WinBUGS before it closes. To do this, add the `noquit` option to `wbsscript`. Now when WinBUGS finds the error, it will stop, place its cursor at the location of the error, and wait. Once you have identified the problem, the WinBUGS window can be closed, the model file corrected, and the script re-created without the `noquit` option.

The OpenBUGS scripting commands are different from those used by the frozen version of WinBUGS, but `wbsscript` knows both versions, so all you need to do to create a script for use with OpenBUGS is to add the `openbugs` option:

```
wbsscript using script.txt, model(model.txt) data(data1.txt+data2.txt) ///
     inits(init.txt) burn(1000) update(5000) thin(5) log(log.txt)     ///
     set(alpha beta) coda(mymod) path(c:\temp) openbugs
```

8.3.5 Running the script

Assuming that the path to WinBUGS has been stored in `executables.txt` and the script file is stored in the current working directory, a WinBUGS script can be run from within Stata using the `wbsrun` command. Then all that is required is the following Stata command:

```
wbsrun using script.txt
```

If the path to WinBUGS has not been written to `executables.txt`, it can be specified using the `executable()` option; see the `wbsrun` help file for details. If you are running the OpenBUGS version of the software, the command must be changed to the following:

```
wbsrun using script.txt, openbugs
```

When the program runs, a DOS command window will open, and Stata will freeze while it waits for WinBUGS to finish. Once the run is completed, the Command window will close automatically, and Stata will become active again. It is possible to have WinBUGS run in the background while Stata is used for other tasks by specifying the `background` option to `wbsrun`, although, of course, this would rarely be sensible if `wbsrun` were called from within a do-file.

8.3.6 Reading the results into Stata

WinBUGS writes the results for each chain into a text file with two values per line, a simulation number and the simulated value of the parameter. The parameters are stored one after another, so if two parameters, `alpha` and `beta`, are to be stored in the same file, it will contain all the simulated values of `alpha` followed by all of those for beta. To enable the start and finish for any parameter to be located, WinBUGS also creates an index file that contains the first and last line number for each parameter. This file format is inconvenient and is not readily readable by Stata, so the command wbscoda reads these files. In its simplest form, all you need to write is

```
wbscoda using mymod, clear
```

where mymod is the prefix to the output files as specified by the coda option in wbsscript.

The command gets its name from CODA, a program for checking the convergence of MCMC simulations that uses the same file format. wbscoda reads the results into Stata and restructures them as a rectangular array with one column for each parameter; it copes automatically with multiple chains by creating a Stata variable that contains the number of the chain.

There are slight differences between the original version of WinBUGS and the OpenBUGS version in the way that the output files are named, so when the OpenBUGS version is being used, the command must be modified to the following:

```
wbscoda using mymod, clear openbugs
```

There is a convention amongst WinBUGS programmers of using a full stop within parameter names. Thus they might use alpha.sd and beta.sd to represent the standard deviations of alpha and beta. Such parameter names are not allowed as variable names in Stata, so wbscoda converts them to alpha_sd and beta_sd. Similar changes are made to the names of items in a vector of parameters, so a[1] and a[2] become a_1 and a_2 within Stata.

8.3.7 Inspecting the log file

wbsscript includes an option for saving a log file containing a record of the run. Such a log file includes the error message should anything go wrong, and if requested in the script, it will also contain the deviance information criterion, a measure of the fit of the model that is described in more detail in chapter 10. The log file can be viewed after the run using Stata's type command, for instance,

```
type log.txt
```

8.3.8 Reading WinBUGS data files

Occasionally, data may be supplied in the WinBUGS list format, and although this could be used directly by WinBUGS, it would not be easy to read it into Stata for data exploration or use in other analyses. One common instance of this is when one wants to run a WinBUGS example problem. The examples can be viewed by using WinBUGS interactively, and the data can then be copied and pasted into a text file. Because those data will be in list format, there is an ado-file called wbsdecode that will automatically read list-formatted files. For example, one might read the contents of example.txt by using

```
wbsdecode using example.txt, clear
```

By adding the `array` option, one can have the command also read WinBUGS data stored in array format.

8.4 Case study 8: Growth of sea cows

The WinBUGS help system contains several complete examples, one of which will be used to show how a model can be processed entirely from within Stata. The example is a set of 27 measurements of age (years) and length (meters) made on sea cows (dugongs). The data were taken from Ratkowsky (1983), and the WinBUGS analysis fits a nonlinear model to the data with the form and priors,

$$
\begin{aligned}
y_i &\sim N(\mu_i, \tau) \qquad i = 1, \ldots, 27 \\
\mu_i &= \alpha - \beta\gamma^{x_i} \\
\alpha &\sim N(0, 0.000001) \\
\beta &\sim N(0, 0.000001) \\
\gamma &\sim U(0.5, 1) \\
\tau &\sim \text{gamma}(0.001, 0.001)
\end{aligned}
$$

where y represents the lengths in meters, and x represents the ages of the 27 animals in years. These priors are given using WinBUGS parameterizations, so 0.000001 is a precision, and the gamma prior has a mean of 1 and corresponds to Stata's gamma(0.001, 1000).

All the examples provided with WinBUGS use vague priors. In this particular example, `alpha` represents the average length of a fully grown dugong measured in meters, so a normal prior with a mean of 0 and a precision of 0.000001 implies that we are confident that `alpha` is something in the range ± 2000 meters. In practice, we might do slightly better than that, but for comparability, we will adopt the priors used in the WinBUGS manual.

The data, which are included in the WinBUGS example files, are reproduced in table 8.1 and displayed in figure 8.1.

Table 8.1. Ages (years) and lengths (meters) of 27 sea cows

Age	1.0	1.5	1.5	1.5	2.5	4.0	5.0	5.0	7.0
Length	1.80	1.85	1.87	1.77	2.02	2.27	2.15	2.26	2.47
Age	8.0	8.5	9.0	9.5	9.5	10.0	12.0	12.0	13.0
Length	2.19	2.26	2.40	2.39	2.41	2.50	2.32	2.32	2.43
Age	13.0	14.5	15.5	15.5	16.5	17.0	22.5	29.0	31.5
Length	2.47	2.56	2.65	2.47	2.64	2.56	2.70	2.72	2.57

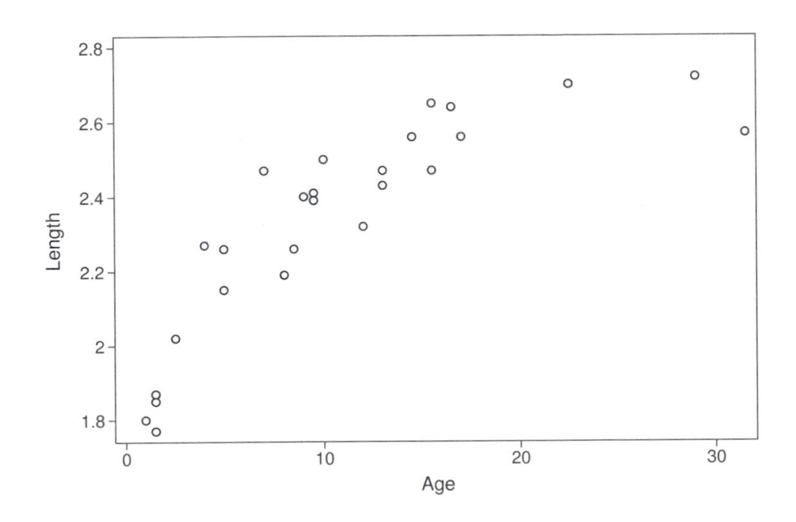

Figure 8.1. Ages and lengths of 27 sea cows

The model file can be prepared in the Do-file Editor and saved as `model.txt`. Its form will be

begin: dugong model file

```
model {
  for (i in 1:27) {
     y[i] ~ dnorm(mu[i],tau)
     mu[i] <- alpha - beta*pow(gamma,x[i])
  }
  alpha ~ dnorm(0.0, 1.0E-6)
  beta  ~ dnorm(0.0, 1.0E-6)
  gamma ~ dunif(0.5,1)
  tau   ~ dgamma(0.001,0.001)
}
```

end: dugong model file

Notice the use of the function `pow()` to create powers and `E` to specify numbers in scientific notation. The program given in the WinBUGS examples file also includes a line for calculating the standard deviation of the residuals. Remembering that tau will be the precision, the corresponding standard deviation is calculated as `1/sqrt(tau)`. However, it is unnecessary to include this, because the results will be read into Stata for subsequent processing, and provided that `tau` is saved, that calculation can be made within Stata when it is required.

If the data are stored in Stata as two variables, `x` and `y`, a data file in list format is obtained by using

```
wbslist (variable x y) using data.txt, replace
```

The initial values could also be created using the `wbslist` commands. In this case, the four parameters are all scalars, so we write

```
wbslist (alpha=1,beta=1,gamma=0.9,tau=1) using inits.txt, replace
```

Finally, the script file needs to be prepared. `wbsscript` needs to know which program you plan to run; we will use OpenBUGS, but results with WinBUGS would be similar. Assuming a burn-in of length 1,000 and a run of length 10,000, the analysis can be specified as follows:

```
wbsscript using script.txt, model(model.txt) data(data.txt)    ///
     inits(inits.txt)  coda(dugong) set(alpha beta gamma tau)  ///
     burn(1000) update(10000) log(log.txt) replace openbugs
```

To run this analysis, read the results into Stata, summarize them, and make a basic assessment of convergence requires the Stata commands

```
wbsrun using script.txt, openbugs
type log.txt
wbscoda using dugong, clear openbugs
generate sd = 1/sqrt(tau)
mcmcstats alpha beta gamma sd
mcmctrace alpha beta gamma sd
mcmccusum alpha beta gamma sd, ref(1)
```

In this case, the OpenBUGS log file, which is reproduced below, shows that the program ran without any errors.

```
model is syntactically correct
data loaded
model compiled
model is initialized
model is already initialized
model is updating
1000 updates took 0 s
model is updating
10000 updates took 0 s
CODA files written
```

The summary of the results produced by `mcmcstats` is

Parameter	n	mean	sd	sem	median	95% CrI
alpha	10000	2.648	0.070	0.0056	2.642	(2.527, 2.791)
beta	10000	0.971	0.075	0.0023	0.973	(0.815, 1.119)
gamma	10000	0.861	0.033	0.0026	0.864	(0.785, 0.913)
sd	10000	0.099	0.015	0.0002	0.097	(0.075, 0.132)

and the resulting trace and cusum plots are shown as figure 8.2 and figure 8.3. Neither indicates any early drift, and the mixing seems reasonable. One noticeable feature of the cusum plots is the similarity between the graphs for `alpha` and `gamma`. This suggests strong correlation between the two parameters, which can be visualized by a simple scatterplot of the simulations as shown in figure 8.4. As well as affecting interpretation, this correlation could also affect the mixing of the Gibbs sampler, although in this case, this does not appear to have been a major problem.

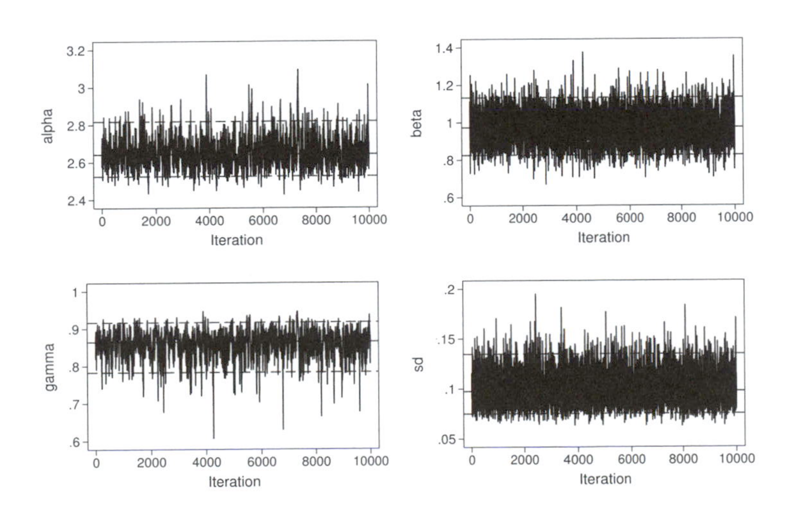

Figure 8.2. Trace plots of the four parameters in the model of the sea cow data

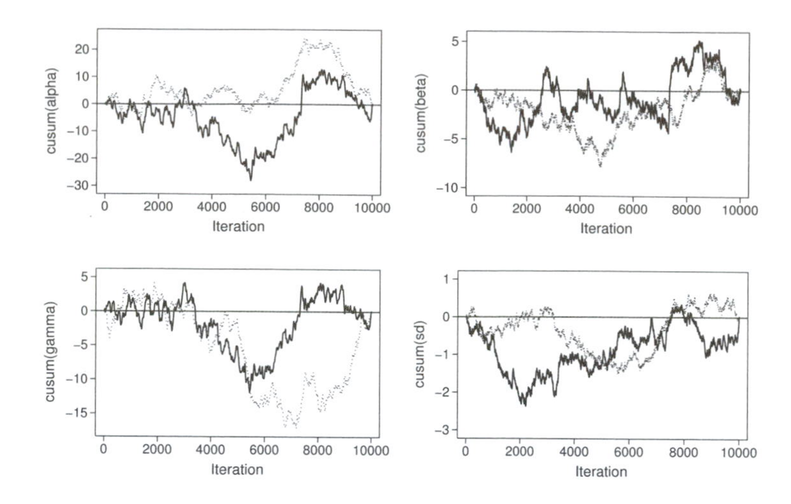

Figure 8.3. Cusum plots of the four parameters in the model of the sea cow data

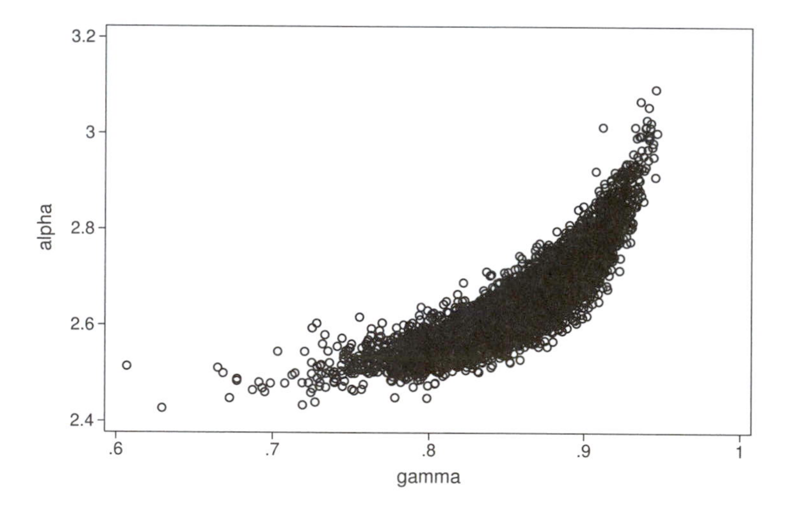

Figure 8.4. Scatterplot of the simulations of `alpha` and `gamma` in the model of the sea cow data

8.4.1 WinBUGS or OpenBUGS

OpenBUGS has been extensively tested to ensure that it gives the same answers as WinBUGS, but the samplers selected by OpenBUGS may well be different from those chosen by WinBUGS. Therefore, although both programs will converge to the same solution, the speed of convergence may be different. When the dugong analysis is rerun in the frozen version of WinBUGS, the estimates that it gives are

```
---------------------------------------------------------------------------
Parameter          n       mean         sd        sem    median          95% CrI
---------------------------------------------------------------------------
 alpha         10000      2.650      0.075     0.0024     2.642 (   2.525,    2.818 )
 beta          10000      0.975      0.077     0.0008     0.973 (   0.825,    1.133 )
 gamma         10000      0.860      0.034     0.0016     0.864 (   0.783,    0.917 )
 sd            10000      0.099      0.015     0.0002     0.097 (   0.075,    0.135 )
---------------------------------------------------------------------------
```

The chain was the same length as that used previously with OpenBUGS, and the initial values were identical, but OpenBUGS took 1.78 sec, while on the same computer, WinBUGS took 2.87 sec, a trivial difference for such a simple model but 60% longer—which if maintained in a more demanding problem could be important. In exercise 2 of chapter 7, the reader was asked to write a Mata program to analyze the same model; it is interesting to note that an efficiently coded Mata program can obtain results comparable to OpenBUGS in about 0.4 sec, although part of the difference is due to the one-off task of loading OpenBUGS.

Both OpenBUGS and WinBUGS appear to be converging to the same solution, but the standard errors for the WinBUGS analysis are smaller, suggesting that its choice of samplers has led to better mixing. We need to be cautious in jumping to conclusions about random processes because part of the difference between programs could just be due to the randomness of the simulations. In this case, however, the difference is consistent when the seed for the random-number generator is changed and seems to reflect a real difference in the programs. The difference in mixing is reflected in figure 8.5, which shows the autocorrelations in the chains produced by WinBUGS and OpenBUGS.

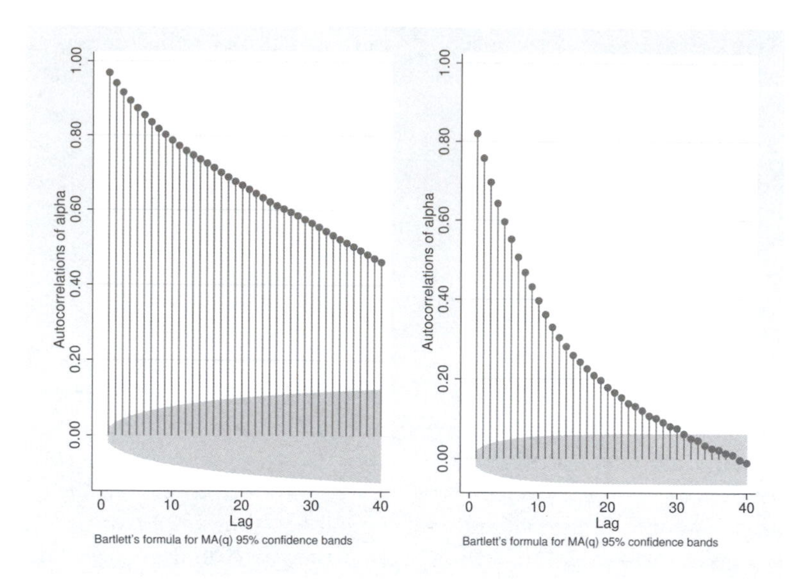

Figure 8.5. Autocorrelation plots of chains parameter `alpha` from the dugong model produced by OpenBUGS (left) and WinBUGS (right)

The differences between OpenBUGS and WinBUGS are model dependent, and so it is not possible to say which program will mix better or run faster. Often, there is a tradeoff; a program that chooses samplers that run quickly may well produce worse mixing. In most cases, the differences in performance between OpenBUGS and WinBUGS seem to be small, but if one of the programs does struggle to converge for a particular model, it may be worth trying the other.

8.5 Case study 9: Jawbone size

The WinBUGS example files contain an analysis of a set of repeated measurements of the jawbone heights of 20 boys taken at 8, 8.5, 9, and 9.5 years of age. These data come from Elston and Grizzle (1962) and are shown in table 8.2 and plotted in figure 8.6. A linear growth model seems reasonable over this relatively short time period, but to capture the variability, one must allow for the correlation between repeated measurements made on the same child. One possible model is to treat the sets of four measurements as coming from a multivariate normal distribution.

Table 8.2. Jawbone heights (mm) of 20 boys measured every 6 months

Age in years			
8	8.5	9	9.5
47.8	48.8	49.0	49.7
46.4	47.3	47.7	48.4
46.3	46.8	47.8	48.5
45.1	45.3	46.1	47.2
47.6	48.5	48.9	49.3
52.5	53.2	53.3	53.7
51.2	53.0	54.3	54.5
49.8	50.0	50.3	52.7
48.1	50.8	52.3	54.4
45.0	47.0	47.3	48.3
51.2	51.4	51.6	51.9
48.5	49.2	53.0	55.5
52.1	52.8	53.7	55.0
48.2	48.9	49.3	49.8
49.6	50.4	51.2	51.8
50.7	51.7	52.7	53.3
47.2	47.7	48.4	49.5
53.3	54.6	55.1	55.3
46.2	47.5	48.1	48.4
46.3	47.6	51.3	51.8

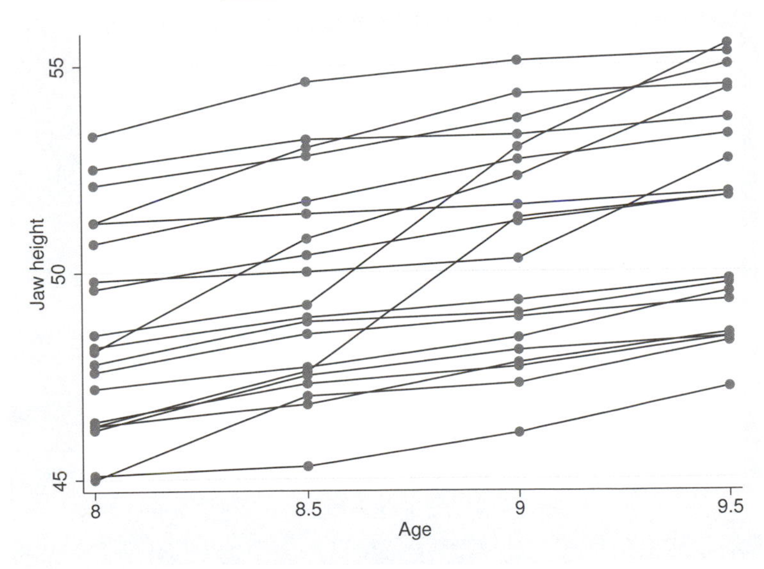

Figure 8.6. Repeated measurements of jaw height on 20 boys taken 6 months apart

WinBUGS parameterizes the multivariate normal in terms of its mean vector and its precision matrix; that is, the inverse of the covariance matrix and the most commonly used prior for a precision matrix is the Wishart. The Wishart distribution has two parameters: a scale matrix, `R`, which must reflect our beliefs about the covariance structure, and a degrees of freedom, `k`, which reflects our confidence in those beliefs. `R` should be chosen to be equal to `k` times a guess at the covariance matrix, and if you wish to make the prior as vague as possible, the degrees of freedom should be set equal to the dimension of the matrix. Together the multivariate normal and Wishart prior produce the following model:

$$
\begin{aligned}
(y_1, y_2, y_3, y_4)' &\sim \text{MVN}(\mu, \Omega) \\
\mu_i &= \beta_0 + \beta_1 * \text{age}_i \\
\beta_0 &\sim N(0, 0.001) \\
\beta_1 &\sim N(0, 0.001) \\
\Omega &\sim \text{Wishart}(\mathbf{R}, 4)
\end{aligned}
$$

In this case, the choice of prior matrix `R` will have little influence on the final answer because the degrees of freedom of the Wishart is set so low.

The model file for this problem is shown below. Notice how the dimensions of the matrices are specified as 1:4, meaning 1 to 4. In fact, the dimensions of the right-hand side of an expression can be omitted because they are implicit in the left-hand side, so it would be equally correct to write `dwish(R[,],4)`.

begin: jaw size model file

```
model {
   for (i in 1:20) {
      y[i,1:4] ~ dmnorm(mu[1:4],Omega[1:4,1:4])
   }
   mu[1] <- beta0 + beta1*8
   mu[2] <- beta0 + beta1*8.5
   mu[3] <- beta0 + beta1*9
   mu[4] <- beta0 + beta1*9.5
   beta0 ~ dnorm(0, .01)
   beta1 ~ dnorm(0.0, 0.001)
   Omega[1:4,1:4]  ~ dwish(R[1:4,1:4],4)
}
```

end: jaw size model file

The way that the model is described requires that the data be supplied a 20×4 matrix called `y`. Assuming that the data are stored in Stata as four variables called `y1`, `y2`, `y3`, and `y4`, each of 20 values, we could either convert the data to a Stata matrix using Stata's **`mkmat`** command and then write it using the matrix keyword with **`wbslist`** or write the data directly by using the **`structure`** option of **`wbslist`**. The matrix `R` also needs to be specified, and this can be done at the same time. So to make `R` equal to the four-dimensional identity and to write it and the data requires

```
matrix R = I(4)
wbslist (matrix R) (structure y1 y2 y3 y4, name(y) using data.txt, replace
```

The WinBUGS manual sets initial values for `beta0` and `beta1` and leaves WinBUGS to generate random initial values for `Omega`. This simply requires

```
wbslist (beta0=40,beta1=0) using inits.txt, replace
```

The script follows the usual pattern, and assuming that OpenBUGS is being used, we could type

```
wbsscript using script.txt, model(model.txt) data(data.txt) inits(inits.txt) ///
        coda(jaw) set(beta0 beta1 Omega) log(log.txt) burn(1000)           ///
        update(10000) replace openbugs
wbsrun using script.txt, openbugs
type log.txt
wbscoda using jaw, clear openbugs
```

By taking the sets of simulated values for `Omega` and inverting the corresponding matrix, we can estimate the posterior distributions of the variances and correlations. Having read the results into Stata, use

begin: inverting the precision

```
matrix OMEGA = J(4,4,0)
forvalues j=1/4 {
   forvalues k=1/4 {
      quietly generate v_`j´_`k´ = .
      quietly generate r_`j´_`k´ = .
   }
}
forvalues i=1/10000 {
   forvalues j=1/4 {
      forvalues k=1/4 {
         matrix OMEGA[`j´,`k´] = Omega_`j´_`k´[`i´]
      }
   }
   matrix V = inv(OMEGA)
   forvalues j=1/4 {
      forvalues k=1/4 {
         quietly replace v_`j´_`k´ = V[`j´,`k´] in `i´
      }
   }
   matrix R = corr(V)
   forvalues j=1/4 {
      forvalues k=1/4 {
         quietly replace r_`j´_`k´ = R[`j´,`k´] in `i´
      }
   }
}
```

end: inverting the precision

The results as given by `mcmcstats` are shown below. Here `v_1_1` is the first term in the variance matrix, and `r_1_2` is the correlation between the first two jaw measurements. The variances increase slightly with age, and the correlations are high and seem to decline roughly geometrically with the interval between measurements.

Parameter	n	mean	sd	sem	median	95% CrI
beta0	10000	32.362	1.969	0.1017	32.470	(28.110, 35.970)
beta1	10000	2.015	0.224	0.0116	2.001	(1.602, 2.491)
v_1_1	10000	6.810	2.437	0.0262	6.301	(3.575, 12.899)
v_2_2	10000	6.925	2.459	0.0262	6.440	(3.648, 13.070)
v_3_3	10000	7.394	2.588	0.0273	6.882	(3.884, 13.837)
v_4_4	10000	7.995	2.828	0.0301	7.400	(4.174, 14.969)
r_1_2	10000	0.960	0.018	0.0002	0.964	(0.916, 0.985)
r_1_3	10000	0.864	0.058	0.0006	0.874	(0.722, 0.946)
r_1_4	10000	0.794	0.085	0.0009	0.808	(0.587, 0.917)
r_2_3	10000	0.914	0.038	0.0004	0.921	(0.818, 0.967)
r_2_4	10000	0.843	0.067	0.0007	0.855	(0.679, 0.938)
r_3_4	10000	0.958	0.020	0.0002	0.962	(0.908, 0.984)

8.5.1 Overrelaxation

In chapter 3, we considered the problems that can arise with samplers that change one parameter at a time in circumstances where two or more of the parameters are highly correlated. At each stage, all other parameters are held fixed, including the parameters that are highly correlated with the parameter being updated, and so there will be little scope for moving the parameter. This often results in a chain that is slow to travel across the posterior. Both WinBUGS and OpenBUGS will block together some of the parameters and update them together in an attempt to reduce this effect, but for situations in which poor mixing persists, they offer an option known as overrelaxation.

Neal (1998) described a simple version of overrelaxation that can be applied to any distribution. Instead of generating a single update for a parameter, the sampler is used repeatedly to give a set of updates. These potential updates are then sorted alongside the previous value of the parameter. If the previous value is the largest of the set then we take the smallest as the chosen update; if it is the second largest, then we take the second smallest update, and so on. In this way, a negative correlation is induced between the previous parameter value and the chosen update, which often leads to improved mixing.

Overrelaxation is not applicable to all samplers; for instance, it is not suitable for conjugate updating or block updating. Where the correlation between parameters is small, there may be no benefit. Even when overrelaxation is beneficial, the improvement is bought at a price; overrelaxation requires the creation of a set of potential updates and so will be slower. There is no way of knowing in advance whether overrelaxation will help, so all one can do is to try it when one comes across a poorly mixing model.

The option `overrelax` of `wbsscript` will ask either WinBUGS or OpenBUGS to use overrelaxation where the chosen samplers allow. The results below show the performance of OpenBUGS for the jaw measurement analysis with and without overrelaxation. The results for WinBUGS are almost identical to those shown for OpenBUGS, suggesting that for this model, the two programs have chosen the same samplers.

```
Without over-relaxation - run time 1.65s
---------------------------------------------------------------------------
Parameter          n      mean        sd       sem    median         95% CrI
---------------------------------------------------------------------------
 beta0         10000    32.362     1.969    0.1017    32.470 (  28.110,  35.970 )
 beta1         10000     2.015     0.224    0.0116     2.001 (   1.602,   2.491 )
---------------------------------------------------------------------------
With over-relaxation - run time 1.87s
---------------------------------------------------------------------------
Parameter          n      mean        sd       sem    median         95% CrI
---------------------------------------------------------------------------
 beta0         10000    32.466     1.946    0.0737    32.520 (  28.530,  36.260 )
 beta1         10000     2.004     0.221    0.0084     1.998 (   1.575,   2.454 )
---------------------------------------------------------------------------
```

Figure 8.7 shows the autocorrelations for the chains produced for the parameter `beta0` with and without overrelaxation. In this model, `beta0` and `beta1` are strongly negatively correlated, but the technique has clearly been effective at reducing the impact of that correlation.

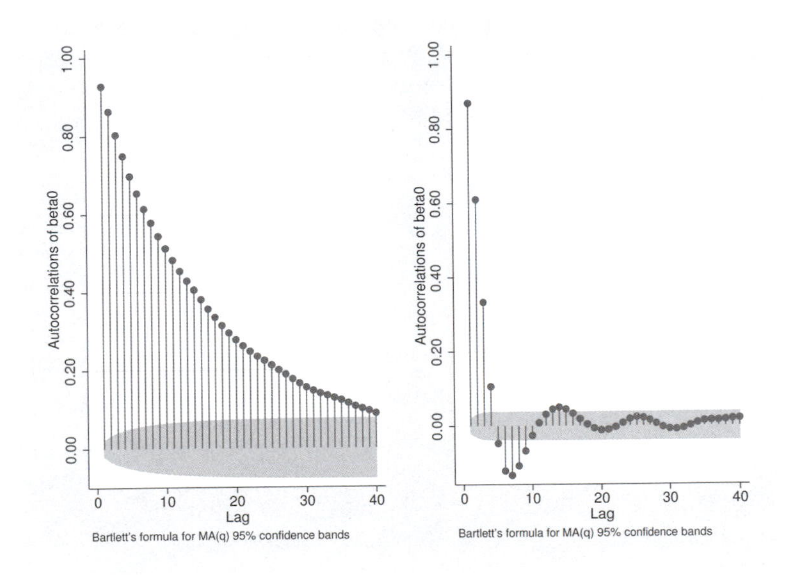

Figure 8.7. Autocorrelation plots of chains produced for the jaw data by OpenBUGS without overrelaxation (left) and with overrelaxation (right)

8.5.2 Changing the seed for the random-number generator

The frozen version of WinBUGS allows the seed of the random-number generator to be changed interactively but does not have a scripting command to enable this to be done within a batch program. Repeat runs of WinBUGS that have the same length therefore produce identical answers. There is a danger in changing the seed of the random-number generator in that different seeds produce sequences of random numbers that go for different lengths before they return to their original state and start reproducing the same sequence. In theory, a long chain could end up duplicating the same random numbers that it produced at the start. To avoid this, OpenBUGS uses a seed that creates an extremely long sequence before it repeats and allows the user to start at 1 of 14 points within the sequence. Each of the predefined starting points is separated by 10^{12} random numbers, which makes it unlikely that there will be duplication within a chain or between two chains started at different points. The starting point can be specified within `wbsscript` by selecting the `seed()` option; for instance, `seed(3)` selects the third starting point. By default, all runs start at the first predefined starting point.

8.6 Advanced features of WinBUGS

8.6.1 Missing data

Missing data are handled automatically by WinBUGS, so all that the user needs to do is to place the missing value symbol, `NA`, in the appropriate places within the data file. All the `wbs` commands for writing data insert `NA` whenever there is a missing value in a Stata variable, so unless you are creating your own data files, there is no need to make any special allowance for missing data.

Suppose that `y` is a vector of 10 data values to be modeled as coming from the same normal distribution but that one of the values, say, `y[5]`, is missing. The model file would describe the distribution as if there were full data:

```
for (i in 1:10) {
     y[i] ~ dnorm(mu,tau)
}
```

During each cycle of the Gibbs sampler, WinBUGS replaces `y[5]` by a new simulated value from a normal distribution with the current values of `mu` and `tau`. Updating `mu` and `tau` subsequently proceeds using the latest simulated `y[5]` as if there were no missing data. If needed, the simulated observations for `y[5]` can be stored using the set option (just as one would save any parameter), and after the run, the posterior distribution of the missing observation can be inspected. The assumption here is that the missingness is not informative; that is, that missing data and observed data come from the same distribution.

8.6.2 Censoring and truncation

With censoring and truncation, there are small differences between OpenBUGS and the frozen version of WinBUGS: both handle censoring in a similar way, but only OpenBUGS has any facilities for handling truncation. Censoring and truncation are handled in similar ways, and so it is important to understand the difference between the two.

Censoring arises when data are collected and, for some subjects, the exact value of the response is unknown, even though the response can still be placed within a range. So no subject is excluded, but for some, we only have partial information about their response. On the other hand, truncation arises when subjects are included in a study only when their observations lie within a specified range. Subjects falling outside the range are excluded completely, and we do not know how many such subjects there are.

The classic example of censoring is a follow-up or survival study that observes the time at which a particular event occurs in each individual. If the follow-up stops at time S, before the event has happened in all the subjects, then there will be some subjects for whom we know only that their event time lies within the range (S, ∞). In WinBUGS, such censored data can be denoted by using the notation `I( , )` together with the distribution that would apply if there were no censoring. Thus

```
x[i] ~ dnorm(mu,tau) I(a,b)
```

means that the observation `x[i]` comes from a normal distribution with mean `mu` and precision `tau` but is only observed to lie in the interval (a,b). At each iteration of the Gibbs sampler, a new simulated value of `x[i]` will be drawn from within the appropriate range. For the sake of backward comparability, OpenBUGS also understands the `I(,)` notation, but the developers have decided that they prefer the equivalent notation `C(,)`, which is not understood by WinBUGS. An event time that occurs after time S could, under a Weibull model, be represented in OpenBUGS as

```
t ~ dweib(r,mu) C(S,)
```

The WinBUGS and OpenBUGS example files consider an experiment in which 80 mice were randomly allocated to one of four treatment groups and their survival time, `t`, observed. For mice that were censored, `t` will contain a missing value, `NA`. The dataset also includes a variable `tcen`, containing the censoring times, or a zero if death was observed within the follow-up period. The variable `g` $= 1, 2, 3, 4$ denotes treatment group. A reasonable model for this analysis written in OpenBUGS syntax is

begin: survival model file

```
model {
  for (i in 1 : 80) {
    t[i] ~ dweib(r, mu[g[i]])C(tcen[i],)
  }
  for (j in 1:4) {
    mu[j] <- exp(beta[j])
    beta[j] ~ dnorm(0.0, 0.001)
  }
  r ~ dexp(0.001)
}
```

end: survival model file

Here the data are modeled by a Weibull distribution with a scale that depends on the treatment group. When the time of death is observed, the censoring time, `tcen`, is zero, so the range of the Weibull is unrestricted; when the time of death is missing because of censoring, `tcen` is set to the censoring time so that a restricted range is used to simulate the missing time. In WinBUGS, the same code would work provided that `C(,)` were replaced by `I(,)`.

Software for handling truncation is a more recent development and is only available in OpenBUGS and then only for a restricted range of the more basic distributions. It is denoted in the model file using the `T(,)` notation, which is similar in structure to that for censoring. Under truncation, we know the exact event times, `t[i]`, but subjects are only observed if their event occurs after time S, so there will be no subjects with times smaller than S. Under a Weibull model, this is represented as follows:

```
t[i] ~ dweib(r,mu) T(S,)
```

8.6.3 Nonstandard likelihoods

Most of the commonly used distributions are available in WinBUGS, but sometimes, one must use a nonstandard distribution such as the truncated Poisson, which was used for nonzero polyp counts in chapter 3. One way to code this is to calculate the likelihood of each observation directly as a set of values, p_i, and then to use a standard distribution that has these values as its likelihood.

Suppose that we construct a variable of the same length as the actual data but that is always equal to 1. If these artificial data were modeled as coming from a Bernoulli distribution with probabilities p_i, then the joint likelihood would be just the product of the p_i's, exactly as it is for the true data under the nonstandard distribution. In the case of the polyp counts, we might have expanded the data to one observation per person and coded the zero-truncated Poisson as follows:

```
C <- 10000
for (i in 1:n) {
   ones[i] <- 1
   L[i] <- pow(lambda,y[i])*exp(-lambda)/(1-exp(-lambda))
   p[i] <- L[i] / C
   ones[i] ~ dbern(p[i])
}
```

Here `C` is a large value of our own choosing. The likelihood `L[i]` is unbounded but probabilities must lie in the range (0,1), so `C` must be a value large enough to ensure the validity of the probabilities. Use of the Bernoulli distribution and a variable set equal to 1 is known as the *ones trick*.

An alternative approach is to calculate the contributions to the log likelihood, `logL[i]`, of each observation under the nonstandard model and then to treat `-logL[i]` as the parameter, μ, of a Poisson distribution for which the artificial observation is zero. Because the probability of a zero observation from a Poisson is $\exp(-\mu)$, the contribution of that observation to the likelihood will become equal to `L[i]`, and the joint likelihood will be the same as it would have been under the nonstandard model. To avoid μ becoming negative, one should add an arbitrary constant, `C`, to all values of `-logL[i]`. This approach is called the *zeros trick*, and for the truncated Poisson, it might be coded as follows:

```
C <- 10000
for (i in 1:n) {
   zeros[i] <- 0
   LogL[i] <- y[i]*log(lambda)-lambda-log(1-exp(-lambda))
   mu[i] <- -logL[i] + C
   zeros[i] ~ dpois(mu[i])
}
```

Usually, there are fewer numerical problems when we calculate the log likelihood, so the zeros trick is often more robust than the ones trick.

In OpenBUGS (but not in WinBUGS), there is a standard distribution called `dloglik()` that simplifies the coding of the zeros trick slightly. Thus the previous example could also be coded as follows:

```
for (i in 1:n) {
   zeros[i] <- 0
   LogL[i] <- y[i]*log(lambda)-lambda-log(1-exp(-lambda))
   zeros[i] ~ dloglik(logL[i])
}
```

8.6.4 Nonstandard priors

A procedure similar to the zeros trick can be used to create nonstandard priors. The parameter is given a flat prior, `dflat()`, and then an artificial zero observation is created and modeled as coming from a Poisson distribution with mean `-log(F[i])`, where `F` is the formula for the density of the nonstandard prior. In OpenBUGS, this could be coded as follows:

```
theta ~ dflat()
zero <- 0
LogL <- log(prior for theta)
zero ~ dloglik(logL)
```

In WinBUGS, the user needs to code the Poisson distribution explicitly.

Unfortunately, this method of programming the prior often leads to a poorly mixing chain, and the method should be used only if it really is impossible to find a standard distribution that approximates the prior sufficiently well. It might even be better to use a histogram approximation to the prior. So if `d[]` is a vector containing the midpoints of the histogram bins and `p[]` is a vector containing the prior probabilities of each bin, then the prior for `theta` could be approximated by coding

```
i ~ dcat(p[])
theta <- d[i]
```

8.6.5 The cut() function

There are rare situations in which it is important to represent data by a model without allowing those data to influence the updating of the model's parameters. As an illustration, suppose that a study was conducted in two stages and that a parameter, θ, is common to both stages. The prior is updated to a posterior using the first-stage data, and then that posterior is used as the prior in the second stage. In the case of a Poisson model with parameter `theta`, a gamma prior, and data `y1` in stage 1 and `y2` in stage 2, this might be programmed as follows:

```
theta ~ dgamma(1,1)
for (i in 1:10) {
   y1[i] ~ dpois(theta)
}
for (i in 1:10) {
   y2[i] ~ dpois(theta)
}
```

In this analysis, it is impossible to recover the posterior from stage 1, because `theta` is updated using both sets of data. To overcome this, we can code the model as follows:

```
theta ~ dgamma(1,1)
for (i in 1:10) {
   y1[i] ~ dpois(theta)
}
ctheta <- cut(theta)
for (i in 1:10) {
   y2[i] ~ dpois(ctheta)
}
```

The `cut()` function stops the data `y2` from influencing the value of `theta`, which will only depend on the first-stage data and the original prior. However, `cut()` does make `ctheta` equal to the current value of `theta`, so information flows from `theta` into the updating of `ctheta`; `theta` will describe the posterior after stage 1, and `ctheta` will describe the posterior after both stages.

8.7 GeoBUGS

GeoBUGS is a project linked to WinBUGS that has developed Bayesian software for fitting models to geographically defined data, such as disease rates in different regions or rainfall measured at different locations. The main features of GeoBUGS are the availability of extra distributions that help describe the correlation between geographical locations and software for plotting maps of the results. Because the plotting can more flexibly be performed in Stata, the only feature that really concerns us is the extra distributions. These distributions have been fully integrated into WinBUGS, so they can be used as if they were part of the main program.

The spatial distributions provided by GeoBUGS include conditional autoregressive (CAR) models for normally distributed data, a robust CAR model based on the Laplace distribution, Kriging models for normal data, and a Poisson model for spatial counts. By viewing a time series as a one-dimensional version of a set of spatial data, one can use these distributions to analyze temporal data. Details of the models and examples of their use are given in the GeoBUGS manual that can be downloaded from http://www.mrc-bsu.cam.ac.uk/bugs/winbugs/geobugs12manual.pdf.

The spatial distributions can be specified in the model file in exactly the same way as any other distribution, so a GeoBUGS model is no different from any other WinBUGS model, and it can be run from within Stata in exactly the same way. Following the analysis, it will usually be important to map the results. This can be done in Stata using the `spmap` command; see http://www.stata.com/support/faqs/graphics/spmap.html.

8.8 Programming a series of Bayesian analyses

One advantage of running WinBUGS from within Stata is that it becomes straightforward to embed Bayesian model fitting within a do-file. In this way, one can run multiple analyses without the need for interactive control. Suppose, for example, that we wanted to investigate the repeated sampling properties of the credible intervals produced by a particular Bayesian analysis. Random datasets could be generated within a Stata loop, written to a data file using **`wbslist`**, and then analyzed in WinBUGS using **`wbsrun`**. Alternatively, it might be important to conduct a sensitivity analysis to investigate the influence of a range of priors on a point estimate. In this case, the model file could be created inside the loop with different priors for each run.

As an illustration of a sensitivity analysis, we will vary the prior for `beta0` in the repeated measures model for the jaw height data considered in case study 9. The WinBUGS manual uses an $N(0, 0.001)$ prior, and this precision of 0.001 is equivalent to a standard deviation of about 30, so with a mean of 0, this prior offers reasonable support for values in the range ± 60 mm but has a preference for values close to 0 mm; for instance, it gives approximately twice the prior probability to values around 20 mm that it gives to values around 40 mm. Vague priors of this type, which take no account of the actual measurements being modeled, can sometimes be more informative than was intended.

To run a sensitivity analysis looking for a different prior for `beta0`, we would fix the script, data, and initial value files and make them identical to those used in the main analysis described earlier. Code for creating the model file can be placed within a Stata loop as shown below.

begin: sensitivity to the prior

```
tempname PF
postfile `PF´ mu sd mn0 mn1 lb0 ub0 lb1 ub1 using results.dta, replace
foreach mu of numlist 0 40 {
 foreach sd of numlist 10(10)100 {
   local precision = 1/(`sd´*`sd´)
   tempname FH
   file open  `FH´ using model.txt, write replace
   file write `FH´ "model {" _n
   file write `FH´ "for (i in 1:20) {" _n
   file write `FH´ "y[i,1:4] ~ dmnorm(mu[1:4],Omega[1:4,1:4])" _n
   file write `FH´ "}" _n
   file write `FH´ "mu[1] <- beta0 + beta1*8" _n
   file write `FH´ "mu[2] <- beta0 + beta1*8.5" _n
   file write `FH´ "mu[3] <- beta0 + beta1*9" _n
   file write `FH´ "mu[4] <- beta0 + beta1*9.5" _n
   file write `FH´ "beta0 ~ dnorm(`mu´, `precision´)" _n
   file write `FH´ "beta1 ~ dnorm(0.0, 0.001)" _n
   file write `FH´ "Omega[1:4,1:4]  ~ dwish(R[1:4,1:4],4)" _n
   file write `FH´ "}" _n
   file close `FH´
   wbsrun using script.txt, openbugs
   wbscoda using jaw, clear openbugs
   mcmcstats beta0 beta1
   post `PF´ (`mu´) (`sd´) (r(mn1)) (r(mn2)) (r(lb1)) (r(ub1)) (r(lb2)) (r(ub2))
 }
}
postclose `PF´
```

end: sensitivity to the prior

This code uses Stata's `file` commands to create a different model file for each prior using combinations of prior means, 0 and 40, and prior standard deviations ranging between 10 and 100. The resulting posterior means of `beta0` and `beta1` are shown together with their 95% credible intervals (CrI) in figure 8.8. In each plot, the x axis shows the standard deviation of the prior, the left-hand estimate of each pair relates to a prior mean of 0, and the right-hand estimate relates to a prior mean of 40. The prior precision chosen in the WinBUGS manual is equivalent to a standard deviation of just over 30 mm, so it does have a small influence on the estimates but probably not enough to make a material difference. It is clear from the plot that any larger precision (smaller standard deviation) would have caused the estimate of `beta0` to be drawn toward the prior mean and that it would also have influenced the estimate of `beta1`. Choosing a more realistic prior mean for `beta0` would make the estimates more robust to the choice of precision.

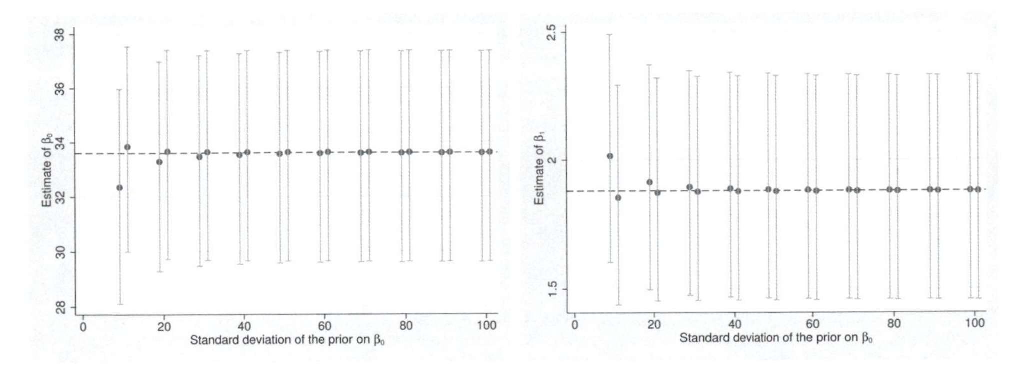

Figure 8.8. Mean and 95% credible intervals for estimates of β_0 and β_1 for different priors on β_0; the left-hand line of each pair relates to a prior mean of 0, and the right-hand line to a prior mean of 40

8.9 OpenBUGS under Linux

Assuming that the Linux version of OpenBUGS has been installed, it should be straightforward to call it from within Stata. In Linux, executables are usually stored together on the search path, but if necessary, the file `executables.txt` must be modified so that the line that begins OpenBUGS contains the path to the Linux executable (remember that Linux is case sensitive). The only places where the operating system makes a difference are to the `wbsscript` command, in which file paths must be written in Linux form, and to the `wbsrun` command, in which the call to the executable must be made using the Linux syntax. Both commands automatically check the operating system when they run, so users do not need to change their code at all. If a Linux or Unix operating system is detected, then the `wbs` commands assume that OpenBUGS is being used, whether or not it is stated explicitly as a command option.

8.10 Debugging WinBUGS

Let's be clear at the outset: debugging WinBUGS code is not easy, and WinBUGS does little to help. If Stata's error messages can be a little vague, then WinBUGS is simply obtuse. Because errors can arise at different stages in the analysis, we need a different approach to correcting each type of error.

The simplest errors to correct are the mistypings and syntax errors that arise when the model file is created. These will be detected when `wbsrun` is used because the script will fail at the first stage when WinBUGS checks the model file. Working from within Stata, you will be aware of a call to WinBUGS because the DOS command window will open, but you will be returned to Stata quickly. The fact that a syntax error was detected will also be evident when the log file is examined. To find out exactly what went wrong, you need to stop the process so that you can see what happened within WinBUGS. The way to do this is to re-create the script with the `noquit` option. Now when the syntax error is detected, WinBUGS will not close down and return you to Stata, but it will stop with its cursor placed at the point in the model file where the error occurred. The error message displayed by WinBUGS will probably not be of any help, so all you can do is search for a syntax error close to the position of the cursor.

Once the model has passed the check for syntax errors, WinBUGS moves on to read the data. If you have prepared the data files using `wbslist` or `wbsarray`, reading the data should not produce any further problems, but if something is wrong, then the `noquit` option can again be used, and WinBUGS will pause with its cursor at the point in the data file where the error was detected. Once again, the user is left to hunt for the source of the problem.

After WinBUGS reads the data, the model is compiled. This is the process whereby WinBUGS turns the model file and the data into code for calculating the conditional distributions of each parameter and then selects the most appropriate samplers. At this stage, any mismatch between the data file and the model file is identified, and this includes finding variables in the data that are not used in the model. Occasionally, with complex models, WinBUGS is unable to find a sampler for some of the parameters. This is one time when it is better to use OpenBUGS because the superior intelligence built into that program means that it is less likely to fail to find a sampler. However, if this does happen, the only options are either to change the model, perhaps by reparameterizing or changing the forms of the prior distributions, or to abandon WinBUGS and use Mata with your own samplers.

After compilation, WinBUGS reads the initial values, and if some parameters are not explicitly initialized, it will try to generate random initial values for those parameters. Occasionally, this can fail, and WinBUGS will be unable to generate initial values that obey the constraints on the ranges of the parameters; in this case, you need to modify your initial values file to specify all the starting values explicitly.

Perhaps the most frustrating errors are those that occur once the MCMC simulations are underway. These generate WinBUGS trap messages that consist of indigestible dumps of everything going on at the time of the failure. The most common cause of error at this stage is numerical overflow; this can be a particular problem for Bayesians who favor ridiculously wide prior distributions. Wide priors can lead the chain to visit some remote and poorly supported regions of the parameter space, especially early in the chain before the regions of high-posterior probability have been identified. If in one of these regions, extreme parameter values need to be exponentiated when calculating the posterior, then it is possible that the numerical results will be too large to be represented within the computer. Alternatively, some term in the calculation might become indistinguishable from zero, and if the calculations require division by that number, the run will fail. Usually, the solution to this problem lies in better initial values and more realistic priors that stop the chain from visiting poorly supported regions of the parameter space. Even if you are wedded to vague priors, it might be better to start with more realistic priors to locate the regions of high-posterior probability. These can then be used to create good initial values for use with the original vague priors. High-dimensional models with flat likelihoods and wide priors will inevitably be prone to these numerical problems, and it can be frustrating to have a program fail after running for several hours. Perhaps this is WinBUGS's way of telling you that you are not using a sensible model.

8.11 Starting points

If you want to read more about WinBUGS, then the best place to start is with the manuals and example files available from the drop-down menus when WinBUGS is run interactively. The WinBUGS and OpenBUGS websites also contain a lot of useful information. Lunn et al. (2000) describe the concepts behind the program and its internal structure, while Ntzoufras (2009) and Lunn et al. (2013) have written comprehensive accounts of WinBUGS.

8.12 Exercises

1. The WinBUGS example files include a set of data on repeat measurements of the industrial yields of samples of dyes taken from six batches. WinBUGS calls this example `DYES`. The data are

Table 8.3. Yields (grams) from repeat samples from one of six batches

Batch	Yields (g)				
1	1545	1440	1440	1520	1580
2	1540	1555	1490	1560	1495
3	1595	1550	1605	1510	1560
4	1445	1440	1595	1465	1545
5	1595	1630	1515	1635	1625
6	1520	1455	1450	1480	1445

Our objective is to compare the within- and between-batch variances using the model

$$\begin{aligned} y_{ij} &\sim N(\mu_i, \tau_w) \\ \mu_i &\sim N(\theta, \tau_b) \end{aligned}$$

where i denotes the batch, and j denotes the repeat measurement within each batch. θ is the overall mean yield, μ_i is the mean yield for batch i, and the τ's are the within- and between-batch precisions.

a. Choose informative priors for the model parameters assuming that it is known in advance that most yields will be between 1,400 g and 1,600 g. Then starting with the data in a Stata dataset (`.dta`), run a Bayesian analysis of this problem from within Stata by calling WinBUGS. Save the chain of values of θ, τ_w, and τ_b, read them into Stata, and check convergence.

b. From your results, estimate the within- and between-batch variances and standard deviations, the total variance (within-batch variance plus between-batch variance), and the intraclass correlation (ICC equals variance between batches divided by total variance). Display the posterior densities of each of these derived statistics, and compare them with the implicit priors that you gave them.

c. Replace your informative priors with vague priors, and see how that affects your estimate of the ICC.

d. The WinBUGS example file includes a second analysis in which priors are placed on θ, τ_w, and the ICC. Place your own informative priors on these quantities, and see how the results compare with your previous informative analysis.

e. Which of the three analyses is preferable?

2. The WinBUGS example files include a set of data called MICE that was referred to briefly in section 8.6.2. Sets of 20 mice were subjected to 1 of 4 treatments and followed up with until either they died or the study ended after 40 weeks, at which time the survival was censored. The manual's suggested model for these data is a Weibull regression in which the scale parameter depends on treatment.

   Use `wbsdecode` to read these data from WinBUGS into Stata. Set your own priors. Prepare the model file and three sets of widely separated initial values. Use WinBUGS to run all three chains at the same time, and read the results into Stata. Assess convergence by comparing the three chains. When you are sure that the chains have converged, estimate the effect of the treatment. Estimate the probability that the average survival time is longer for group 2 than for any of the other groups.

9 Model checking

9.1 Introduction

Proposing a model for an analysis is something of an art: the statistician must select a structure flexible enough to capture both the trend and the random variation in the observations without making it so flexible that the trend in the model follows every random variation in the data. Parameter estimation will identify the best-fitting model from within the proposed structure, but if the initial framework is not flexible enough, even the best fit may not represent the data adequately.

Model checking is the process of assessing whether any aspects of the data lie outside the scope of the proposed model. Usually, this involves two stages: first, investigating whether the fitted model captures the trend in the data so that the model will make good predictions; and second, ensuring that the pattern of random variation about that trend is well described so that the uncertainty in the predictions can be quantified accurately. The usual method of model checking is to compare actual observations with predictions; if the agreement is poor, it is often helpful to look for patterns in the disagreement because these can sometimes suggest ways in which the original model could be modified.

Ideally, different sets of data would be used for model fitting and for model checking, but in practice, this is rarely possible. Thus one must remember that using the same data for fitting and checking can create a conservative process, particularly when the dataset is small.

The principles of model checking are similar, whether the model is fit by Bayesian or non-Bayesian methods. The main advantage that Bayesians have is that their predictions are expressed as predictive distributions, so they have ready access not just to the best prediction but also to a measure of its uncertainty.

9.2 Bayesian residual analysis

Suppose that the statistician's chosen family of models for the response y is $p(y|\theta)$, where θ represents the unknown parameters that have been given a prior distribution, $p(\theta)$. In the following discussion, it is important to distinguish clearly between the observed responses and predictions under the fitted model, so the actual data will be written as y and the predictions as y^*. The first Bayesian method for model checking was suggested by Box (1980), who considered the predictive distribution,

$$p(y^*) = \int p(y^*|\theta)p(\theta)d\theta$$

This distribution averages the proposed model for the data over our prior beliefs about the parameters and so is often referred to as the prior predictive distribution. Comparing actual observations with this predictive distribution will help identify poor models provided that the prior distributions are informative, although poor agreement might result from either an inadequate model or prior beliefs that turned out to be unrealistic.

When the priors are vague, it is unlikely that any observation will be described as unexpected in comparison with its prior predictive distribution, so Rubin (1984) suggested using the predictive distribution obtained from the posterior of θ rather than its prior. This produces the posterior predictive distribution,

$$p(y^*|y) = \int p(y^*|\theta)p(\theta|y)d\theta$$

This approach to prediction is especially well suited for use in a Markov chain Monte Carlo (MCMC) analysis because at iteration t, the current simulated parameter estimate $\theta^{(t)}$ can be used to generate a random prediction, y^*, from the distribution $p(y^*|\theta^{(t)})$. Over the full chain, these random predictions describe the posterior predictive distribution.

Rubin's (1984) posterior approach is often a useful basis for model checking, but it is not without its own problems. The observed data are used to learn about θ, and then the estimated values of θ are used to generate the predictions. Because the real data are used both to guide the creation of the predictions and then to assess the model fit, it is not surprising that this approach can be conservative. With very large datasets, it is possible to use part of the data for model fitting and the remainder for assessing fit. Alternatively, we could leave out one observation at a time and refit the model using the remainder. These cross-validation techniques might be ideal, but they are unlikely to be practical. Marshall and Spiegelhalter (2003) proposed what they called a mixed replication. The parameters without hierarchical structure, loosely equivalent to fixed effects, are unlikely to be heavily influenced by individual observations because they are typically estimated from contrasts of averages. On the other hand, random effects and parameters with hierarchical structure can be highly dependent on a few observations. Marshall and Spiegelhalter (2003) suggested that when simulating y^* during an MCMC run, the current values of the fixed parameters should be used but that new values should be generated for the random effects.

According to a mixed replication, predictions from a random-effects logistic regression,

$$\begin{aligned} y_i &\sim B(p_i, n_i) \\ \text{logit}(p_i) &= \beta_0 + \beta_1 x_i + u_i \\ u_i &\sim N(0, \sigma) \end{aligned}$$

would be obtained by first randomly generating a new set of random effects, u_i^*, from a normal distribution with mean zero and standard deviation equal to the current value of σ. Using these new random effects and the current β_0 and β_1, we can find the values of p_i^* and generate predictions y_i^* from the appropriate binomial distribution; that is,

$$\begin{aligned} u_i^* &\sim N(0, \sigma^{(t)}) \\ \text{logit}(p_i^*) &= \beta_0^{(t)} + \beta_1^{(t)} x_i + u_i^* \\ y_i^* &\sim B(p_i^*, n_i) \end{aligned}$$

This approach is easily programmed into Mata or Stata or added to a WinBUGS model file. It works well in most situations, but one must take care when the model contains poorly estimated fixed effects, as in a problem with many unrelated strata, each with few observations; estimates of fixed effects for such strata could be highly dependent on a few data points, and thus cause the residual analysis to be conservative.

Each observed response, y_i, can be compared with the whole chain of simulated predictions, y_i^*, to see whether the observed data are unusual in some sense. This comparison might be captured by calculating any one of several measures of surprise, including the posterior predictive p-value,

$$p_i = \Pr(y_i^* > y_i)$$

the two-tail posterior predictive p-value,

$$\text{tp}_i = 2 \min(p_i, 1 - p_i)$$

the relative predictive surprise,

$$\text{RPS}_i = \frac{p(y_i)}{\max\ p(y_i^*)}$$

or the standardized residual, based on the mean and standard deviation, sd, of the predictive distribution,

$$r_i = \frac{y_i - \text{mean}\{y_i^*\}}{\text{sd}\{y_i^*\}}$$

As with frequentist model checking, it is sometimes helpful to define a function of the data that captures some aspect of the fit that is of particular interest. If T is such a function, then the value of $T(y)$ can be compared with the corresponding simulated predictions, $T(y_i^*)$, using any one of the measures of surprise.

9.3 The mcmccheck command

`mcmccheck` is a downloadable Stata program that implements graphical Bayesian residual checking. As input, it requires the original data and a sample of values from an appropriate predictive distribution. Given this information, `mcmccheck` will calculate the various measures of surprise and plot them in the ways listed in table 9.1. These plots are illustrated in the following sections.

Table 9.1. Types of model-checking plot available through `mcmccheck`

Plot	Description
Predictive distribution	
histogram	histogram of a predictive distribution
density	smoothed predictive density estimate
Measures of surprise	
scatter	scatterplot, e.g., measure of surprise by fitted values
lowess	scatterplot plus a lowess smooth line
epp	empirical probability plot of a measure of surprise
box plot	box plot of a measure of surprise
summary	combination of scatter, epp, and histogram

9.4 Case study 10: Models for Salmonella assays

The Ames *Salmonella*/microsome assay can be used to detect substances that are potentially carcinogenic. A strain of *Salmonella* bacteria that requires histidine for growth is placed on agar containing only trace amounts of histidine. The bacteria grow slowly, but occasional genetic mutations arise, and eventually, some bacteria revert to a form that no longer requires histidine; such cells flourish and form clearly visible colonies. If the *Salmonella* bacteria are treated with a substance that increases the rate of genetic mutation, then more of these histidine-independent colonies are seen.

The Ames *Salmonella*/microsome assay has been widely used, and several authors have discussed methods for analyzing study results. In an early article, Margolin, Kaplan, and Zeiger (1981) criticized the use of Poisson models, and to illustrate their point, they presented an example in which 18 plates were treated with 1 of 6 doses of quinoline. These data are reproduced in table 9.2.

Table 9.2. Number of revertant colonies in the quinoline study reported by Margolin, Kaplan, Zeiger (1981)

Dose	Replicates		
0	15	21	29
10	16	18	21
33	16	26	33
100	27	41	60
333	33	38	41
1000	20	27	42

Breslow (1984), Lawless (1987), Saha and Paul (2005), and the WinBUGS manual have all used the quinoline assay of Margolin et al. (1981) to illustrate methods of *Salmonella* assay analysis. Suggested methods include the use of an overdispersed Poisson regression and models based on the negative binomial.

The analysis in the WinBUGS manual uses a generalized linear mixed model with a Poisson distribution, a log link, and a linear predictor, η, given by

$$\eta_{ij} = \beta_0 + \beta_1 \log(10 + \texttt{dose}_i) + \beta_2 \texttt{dose}_i/100 + u_{ij}$$

where $i = 1, \ldots, 6$ denotes doses, and $j = 1, \ldots, 3$ denotes plates within doses. The random effect, u, is assumed to be normally distributed with mean zero and constant variance.

To compare the model checking plots with those for a less well-fitting model, we will also fit the model without the term in log dose; that is,

$$\eta_{ij} = \beta_0 + \beta_2 \texttt{dose}_i/100 + u_{ij}$$

9.4.1 Generating the predictions in WinBUGS

The following model file is written both to fit the generalized linear mixed model and to approximate the posterior predictive distribution using a mixed replication. The priors used are those suggested in the WinBUGS example file.

begin: model file

```
model {
  for (i in 1 : 18) {
     # MODEL FITTING
     y[i] ~ dpois(mu[i])
     log(mu[i]) <- beta0 + beta1*x1[i] + beta2*x2[i] + u[i]
     u[i] ~ dnorm(0.0, tau)
     # PREDICTIONS
     v[i] ~ dnorm(0.0, tau)
     log(m[i]) <- beta0 + beta1*x1[i] + beta2*x2[i] + v[i]
     ystar[i] ~ dpois(m[i])
   }
   beta0 ~ dnorm(0.0,1.0E-6)
   beta1 ~ dnorm(0.0,1.0E-6)
   beta2 ~ dnorm(0.0,1.0E-6)
   tau ~ dgamma(1.0E-3, 1.0E-3)
}
```

end: model file

At each iteration, new random effects, `v`, are generated, and these are used with the current estimates of the `beta`'s to find the predicted means, `m`, and then to create predicted observations, `ystar`.

The initial values file should contain the single line

```
list(beta0=0, beta1=0, beta2=0, tau=1)
```

The Stata code required to fit this model and to save the information needed for model checking is given below. The original data are saved in the file `salm.dta`, and the parameter estimates and the predictions, `ystar`, are stored in the file `mcmc.dta`. WinBUGS stores the predicted values as `ystar[1]`, `ystar[2]`, ..., which when read by `wbscoda` become `ystar_1`, `ystar_2`, ...

begin: running WinBUGS

```
use salm.dta, clear
generate x1 = log(x+10)
generate x2 = x/100
save salmplus.dta, replace
summarize x1
local m1 = r(mean)
replace x1 = x1 - r(mean)
summarize x2
local m2 = r(mean)
replace x2 = x2 - r(mean)
wbslist (vector x1 x2 y) using data.txt, replace
wbslist (beta0=0,beta1=0,beta2=0,tau=1) using inits.txt, replace
wbsscript using script.txt, ///
   model(model.txt) data(data.txt) init(inits.txt) ///
   burn(1000) update(10000) set(beta0 beta1 beta2 tau u ystar) ///
   coda(sal) replace openbugs
wbsrun using script.txt, openbugs time
wbscoda using sal, clear openbugs
replace beta0 = beta0 - `m1'*beta1 - `m2'*beta2
generate sd = 1/sqrt(tau)
save mcmc.dta, replace
```

end: running WinBUGS

Similar code can be used to fit the model without the term in log dose.

Equivalent predictions could also be generated using Stata or Mata, but we will leave discussion of those methods until sections 9.5 and 9.6 to concentrate on the process of model checking.

9.4.2 Plotting the predictive distributions

There are relatively few observations in the *Salmonella* dataset, so model checking can only be expected to identify gross errors, but we start with the most basic assessment of fit, which is to take a single observation and to compare it with its predictive distribution. Because the approximate predictive distribution is represented by a sample of values, such a comparison might use either a histogram or a smoothed density derived from the predictions.

Showing the observed value against the background of the predictive distribution gives a pictorial representation of the measures of surprise and highlights any inconsistency that would suggest a poorly fitting model. For illustration, the predictive distribution of the observation that looks likely to be the most extreme (the value 60 obtained with a dose of 100) is shown in figure 9.1. It was created using the Stata command

```
mcmccheck, d(y[12]) p(ystar_12) df(salm.dta) pf(mcmc.dta)
```

where `salm.dta` is the file containing the raw data, and `mcmc.dta` is the file containing the MCMC simulations and the predictions. These datasets are specified in options called `dfilename()` and `pfilename()`, respectively, which have been abbreviated to `df()` and `pf()`. The `data()` option is abbreviated to `d()` and specifies the value of the original observation, while the `predictions()` option, abbreviated to `p()`, gives the name of the variable containing the values from the predictive distribution. When the `data()` option refers to a single number, the default plot is a histogram.

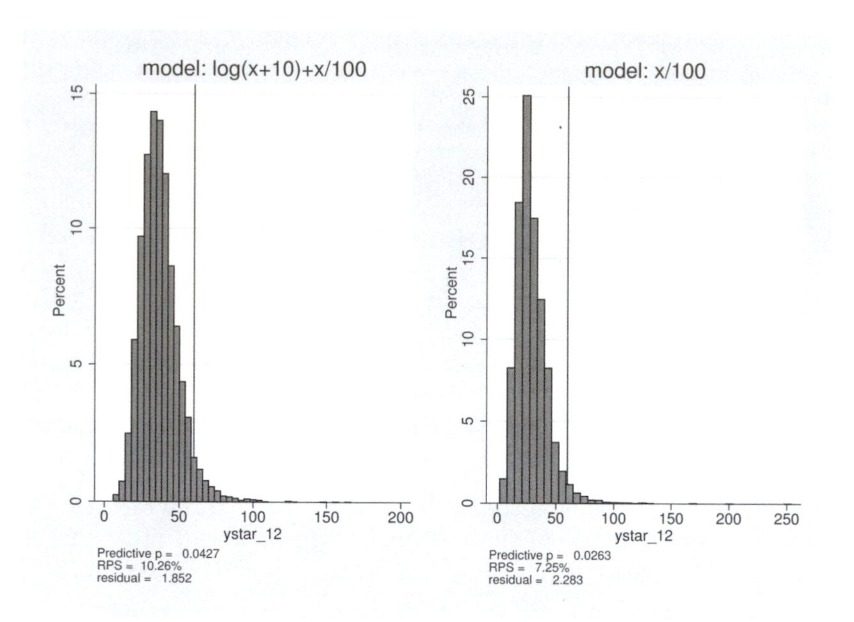

Figure 9.1. *Salmonella* assay: Predictive distributions of the twelfth observation under the two models

As the note below the left-hand histogram shows, under the model suggested by WinBUGS, the observed value of 60 has a standardized residual of 1.85, a predictive p-value of 0.043, and a relative predictive surprise of 10%. All of these measures confirm that the observed value is high but consistent with its predictive distribution. With the poorer-fitting model, all of these measures become slightly more extreme, giving more cause for concern, although probably not enough in themselves to make us seriously question the model given that we chose to look at this observation because it seemed to be the most unusual.

A histogram is the default plot when a single observation is plotted against its predictive distribution. The alternative is to plot the predictive distribution as a smoothed density as plotted by Stata's `kdensity` command. This plot is not shown here because it is similar to the histogram. It could be obtained by using

```
mcmccheck, d(y[12]) p(ystar_12) df(salm.dta) pf(mcmc.dta) plot(density)
```

9.4.3 Residual plots

In non-Bayesian statistics, a scatterplot of residuals against fitted values is often used to identify a pattern of poor fit. The Bayesian equivalent is to plot one of the measures of surprise against the corresponding average predicted value. A plot of Bayesian standardized residuals against mean predictions will work well unless the distribution of the mean predictions is skewed. In this case, the pattern in the plot can be difficult to interpret, and it may be preferable to base the plot on a transformation of the mean prediction, such as its log. When the model fits well, such a residual plot shows standardized residuals across the whole range of predictions that lie within the approximate range ± 2. Single large residuals indicate individual observations that are poorly predicted by the model, while a trend in the residuals would lead us to question whether the proposed model can capture the trend in the data. Conversely, a pattern in which the residuals are more variable in some parts of the range than in others indicates that the model has failed to capture the random variation adequately. Often, problems of variability are clearer if the absolute values of the residuals are plotted against the mean predictions.

The basic scatterplot of residuals against the mean predicted (fitted) values can be created by using the command

```
mcmccheck, d(y) p(ystar_) df(salm.dta) pf(mcmc.dta) x(fit) y(resid)
```

and the plot that is generated is shown in figure 9.2. Because the `data()` option refers to a variable, we assume that there is a set of predictions for each observation called `ystar_1`, `ystar_2`, ... The plot for the model suggested by WinBUGS looks reasonable, but the plot for the model that is linear in dose would probably cause us to doubt its appropriateness.

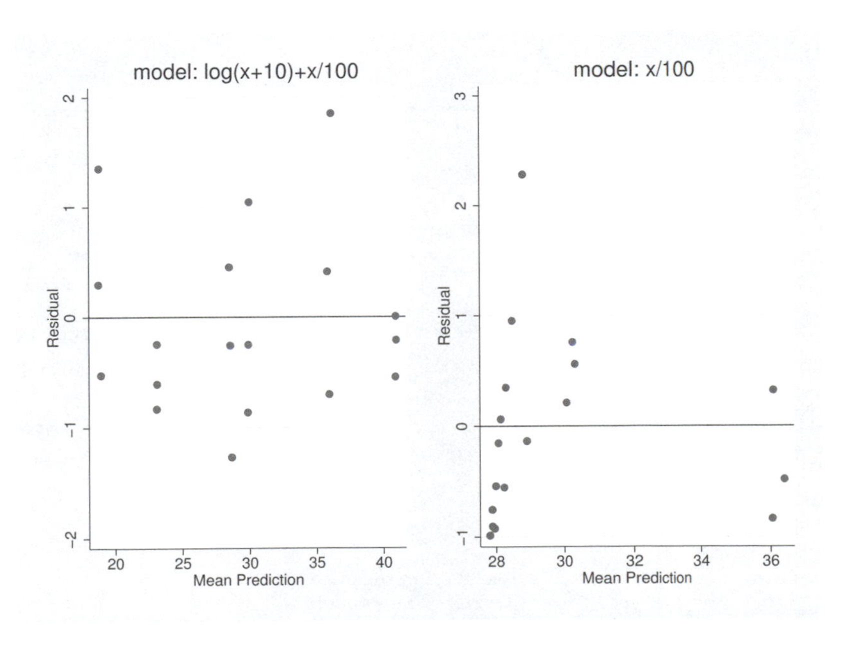

Figure 9.2. *Salmonella* assay: Standardized residuals against the mean of the predictive distribution under the two models

The fitted values need careful interpretation because the model suggested by WinBUGS is not monotonic in dose, x, as can be seen in figure 9.3, which shows the means of the simulated predictive distribution against log dose and was created using the command

```
mcmccheck, d(y) p(ystar_) df(salm.dta) pf(mcmc.dta) x(log(x+10)) y(fit)
```

The predictive distributions for the three plates with the same dose should be the same and differ only because of the sampling inherent in an MCMC analysis.

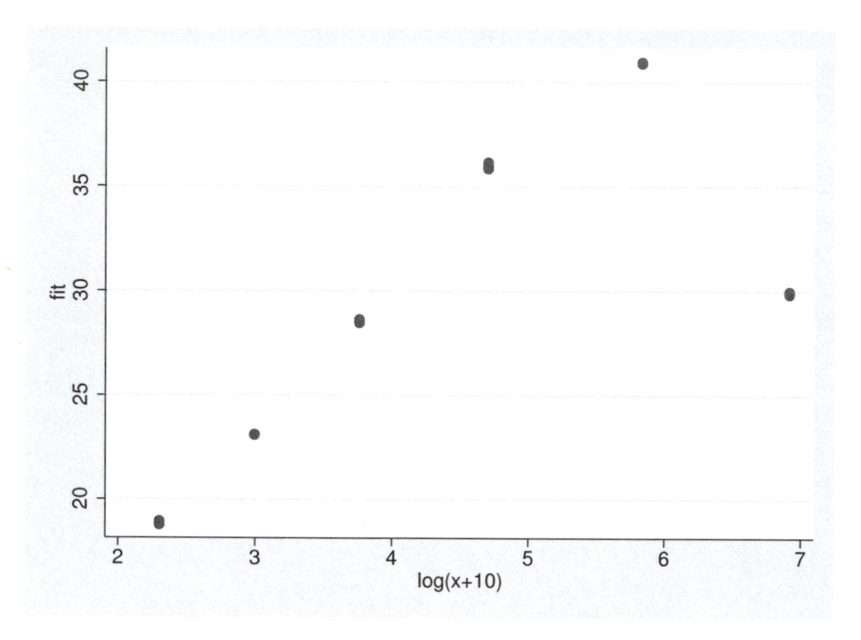

Figure 9.3. *Salmonella* assay: Average predicted values plotted against log(dose+10) for the model suggested by WinBUGS

Whichever measure of surprise is plotted, they are all calculated by `mcmccheck`, and you can save all of them in a Stata dataset. This is useful if you want to inspect them to identify particular points or to use them to create your own plot. The saved dataset contains the standardized residuals stored under the name `res`, the absolute residuals (`absres`), the mean predictions (`fit`), the posterior predictive p-values (`ppp`), twice the posterior predictive p-values (`tppp`), and the relative predictive surprises (`rps`). To plot residuals against log(x+10) and save the output requires

```
mcmccheck, d(y) p(ystar_) df(salm.dta) pf(mcmc.dta) ///
     (log(x+10)) y(res) save(salmres.dta, replace)
```

In figure 9.4, the posterior predictive p-values for the model that is linear in dose are plotted and labeled by the variable `id` from the original dataset. The plot can be produced using the commands

```
use salm.dta, clear
generate id = _n
```

```
save salmid.dta, replace
mcmccheck, d(y) p(ystar_) df(salmid.dta) pf(mcmc.dta) ///
     x(log(x+10)) y(ppp) plot(scatter) lv(id)
```

In this code, the variable `id` is added to the dataset to indicate the observation number, and that variable is used to label the points on the scatterplot using the `lvariable()` option. The pattern of residuals for the poorer-fitting model that is linear in dose demonstrates that observation 12 was indeed the most extreme.

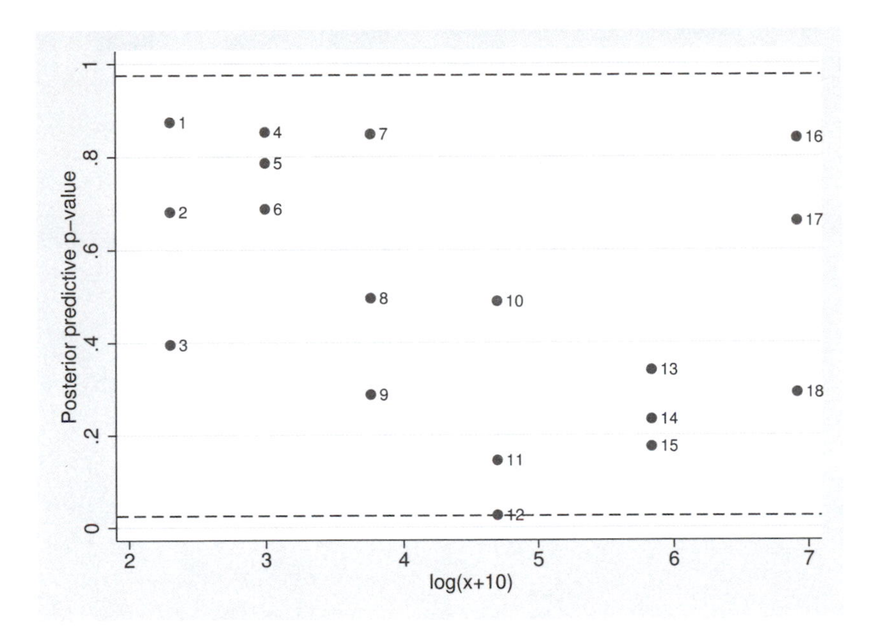

Figure 9.4. *Salmonella* assay: Posterior predictive p-values plotted against log(dose+10) for the poorer-fitting model that is linear in dose

When there is an important categorical covariate, a box plot is a useful way of investigating the pattern in the residuals. Even when the main covariate is continuous, it is sometimes possible to see patterns more clearly when that covariate is categorized. In the *Salmonella* analysis, the treatment variable, `dose`, has six levels, but there are only three replicates at each dose, so a box plot is not ideal. Nevertheless, for illustration, a box plot of the relative predictive surprise (`rps`) is shown in figure 9.5. It was created using the command

```
mcmccheck, d(y) p(ystar_) df(salm.dta) pf(mcmc.dta) ///
     y(rps) plot(box) gopt(over(x) b1title(Dose) ytitle(Relative Surprise))
```

Here `gopt()` is an abbreviation for `goptions()`, which is an option that can be used with any of `mcmccheck`'s plots. Anything included in the option will be passed directly to Stata's `graph` command, and here we take advantage of the fact that `graph box` allows the option called `over()`. Other graphics options are used to label the axes.

Because of the curvature in the response curve, the model that is linear in dose cannot accurately predict the response for doses in the middle of the range.

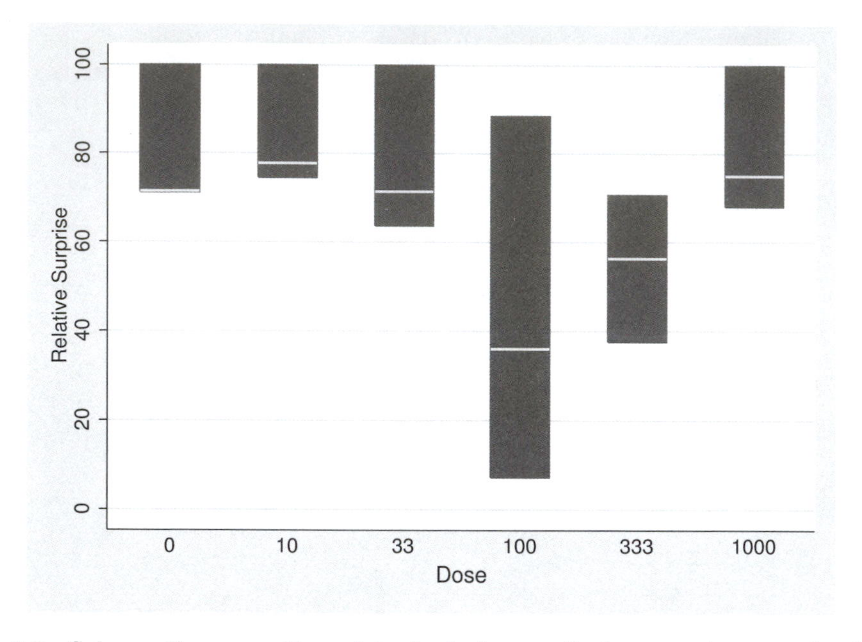

Figure 9.5. *Salmonella* assay: Box plot of relative predictive surprise over the dose for the poorer-fitting model that is linear in dose

9.4.4 Empirical probability plots

Problems with the distributional assumptions of the model can sometimes be identified by making a probability plot, that is, a plot of sorted residuals against the expected ordered residuals under the proposed model. Finding the expected value of the largest residual from a given distribution is often theoretically difficult, but fortunately, the MCMC sample can be used to provide empirical estimates of these order statistics and so enable us to draw an empirical probability plot.

The set of predictions made at any step of the MCMC process can be thought of as a random set of data generated under the fitted model. Ideally, one would like to refit the model to each set of predictions to obtain random residuals, but this is impractical. Instead, we compare the individual predictions with the estimated predictive distributions based on all MCMC iterations to generate the residuals or any other measure of surprise. Ordering the random set of residuals provides a typical largest residual under the model. By repeating this process for different random sets of data chosen from within the MCMC chain, we obtain not just the expected ordered residuals but also an indication of the range within which we would expect the ordered residuals to lie. Plotting observed residuals against expected residuals and indicating the 95% range give the empirical probability plot. Such a plot is shown in figure 9.6 and was created with the command

```
mcmccheck, d(y) p(ystar_) df(salm.dta) pf(mcmc.dta) yaxis(resid)          ///
    plot(epp) nsamples(50)                                                 ///
    gopt(title(model: log(x+10)+x/100) yscale(range(-3 6)) ylabel(-2(2)6))
```

In this case, the expected values were obtained by taking 50 random datasets from within the chain. The distributional assumptions seem reasonable for both models for the *Salmonella* data; the observed residuals lie close to their expected values and well within the limits of the variability.

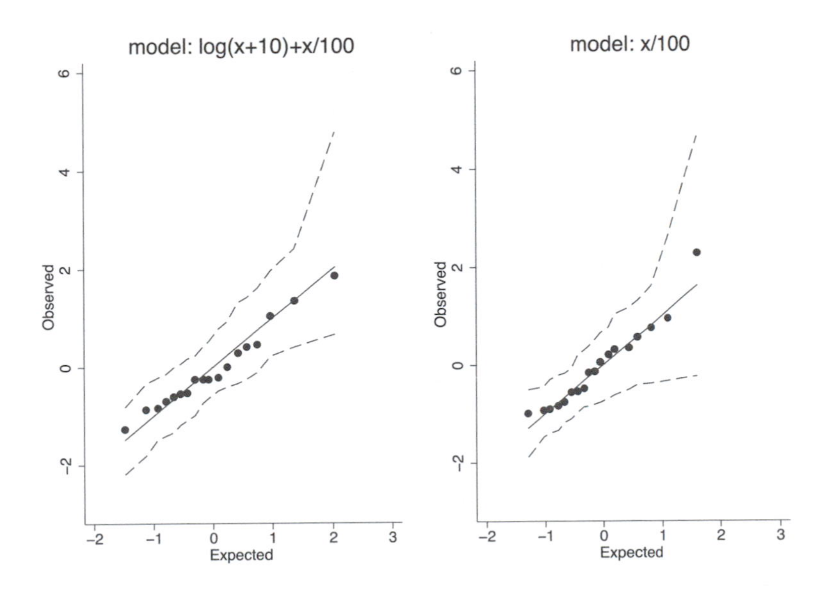

Figure 9.6. *Salmonella* assay: Empirical probability plots of the residuals for the two models

To illustrate the empirical probability plot's ability to detect a misspecified model, we simulated two datasets under one model and then analyzed them under another. In the first example, 40 observations were generated from a log-normal distribution using the command `gen y=exp(rnormal())` and the random-number seed 275904; these data were then analyzed using a normal model with unknown mean and variance. The left-hand part of figure 9.7 shows a histogram of the simulated data, and the right-hand part of the figure shows the empirical probability plot obtained from the Bayesian analysis. The short tail to the left and long tail to the right are evident in both plots. In the second example, 40 overdispersed Poisson observations were simulated with the same seed by first drawing the 40 Poisson means from a gamma distribution using `gen mu=rgamma(2,2.5)` and then drawing the actual data from Poisson distributions with those means using `generate y=rpoisson(mu)`. The data were analyzed as if they came from a standard Poisson distribution with unknown mean and no overdispersion. Figure 9.8 shows the histogram of the simulated data and the empirical probability plot obtained from the Bayesian analysis. The overdispersion can be seen in the curving of the probability plot and the excess of large values in the histogram, although the latter is less easy to see.

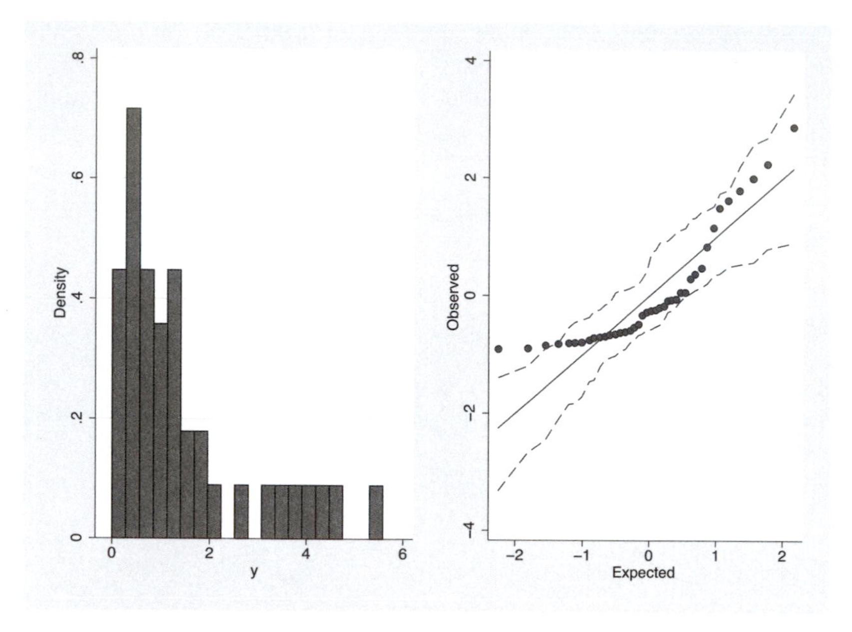

Figure 9.7. Histogram of the log-normal data and an empirical probability plot of the residuals when the data are analyzed using a normal model

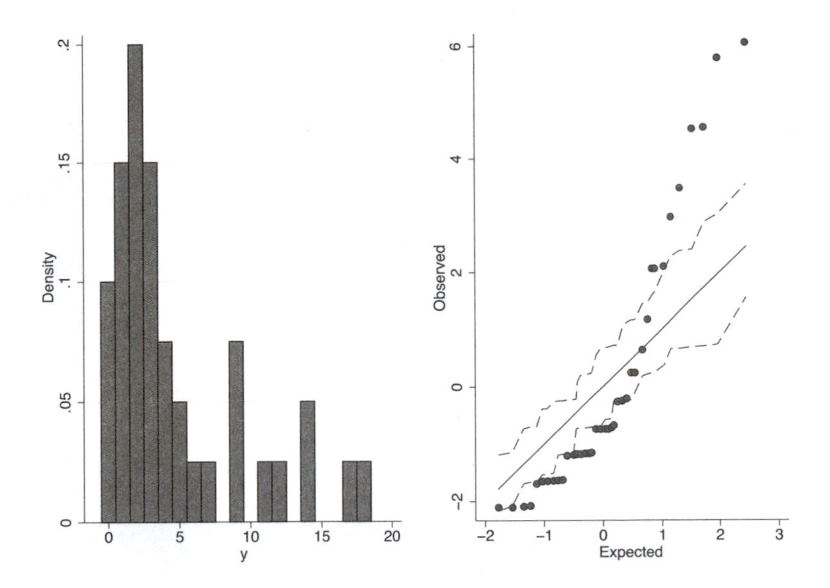

Figure 9.8. Histogram of the overdispersed Poisson data and an empirical probability plot of the residuals when the data are analyzed using a standard Poisson model

9.4.5 A summary plot

`mcmccheck` can also produce a summary residual plot similar to the plot suggested in the book by Lee, Nelder, and Pawitan (2006) and in several articles by the same authors. It is produced when `summary` is specified in the `plot()` option. For example, the summary plot for the model suggested by WinBUGS is shown in figure 9.9 and was produced by typing

```
mcmccheck, d(y) p(ystar_) df(salm.dta) pf(mcmc.dta) plot(summary)
```

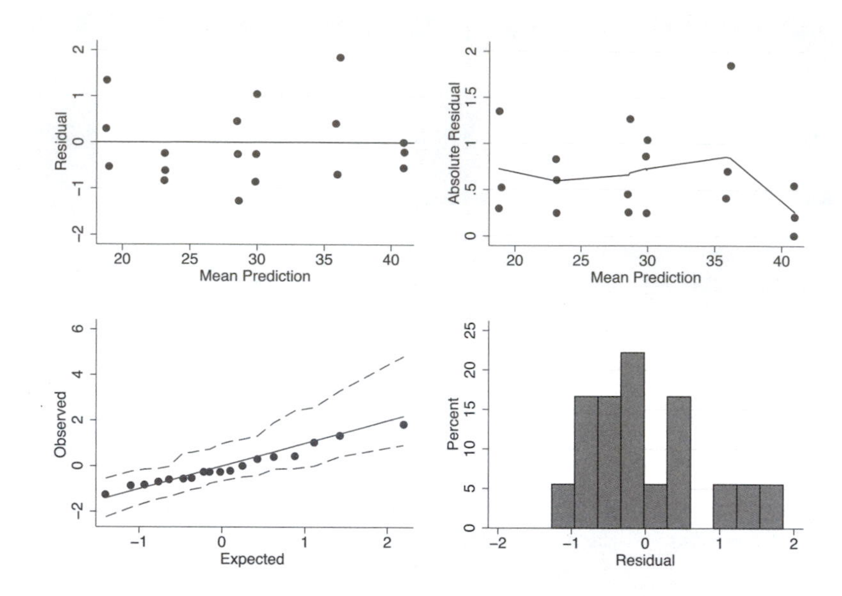

Figure 9.9. *Salmonella* assay: Summary residual plot for the model suggested by WinBUGS

The summary plot for the model that is linear in dose is shown in figure 9.10.

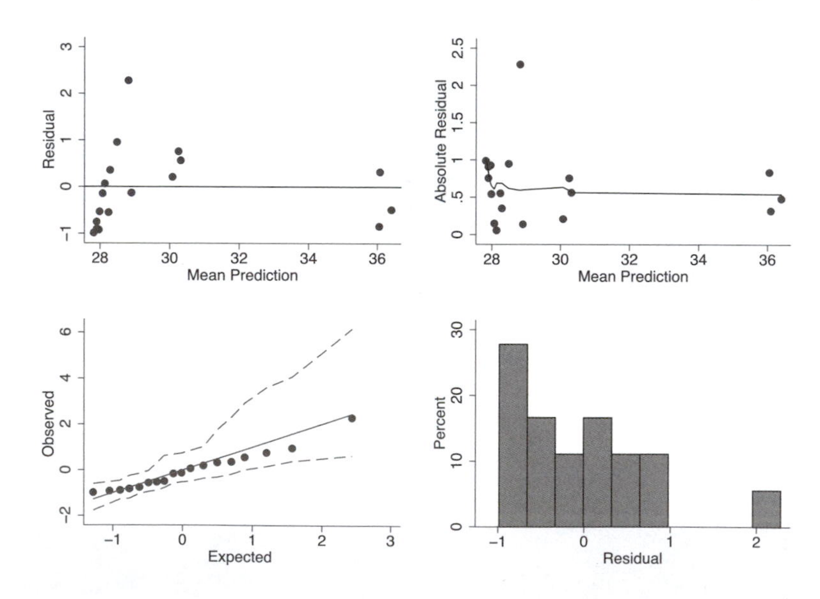

Figure 9.10. *Salmonella* assay: Summary residual plot for the model that is linear in dose

The plots consist of four parts. The top left-hand plot shows residuals against mean predictions, which is useful for identifying outliers and nonlinearity. The top right-hand plot shows absolute residuals against mean predictions together with a lowess smoothed trend line. This plot is useful for detecting problems with the variance structure. The empirical probability plot in the bottom left-hand corner helps diagnose problems with the assumed distribution, and the histogram summarizes the shape of the pattern of residuals. The plot is a good starting point for checking any model and in this example shows that there is no reason to doubt the appropriateness of the model used by WinBUGS, while the model that is linear in dose fits much more poorly.

9.5 Residual checking with Stata

To enable the checking of a Bayesian model fit with Stata, we need a little extra code to calculate the predictions. The script below analyzes the *Salmonella* data using Metropolis–Hastings samplers programmed in Stata. It starts and ends in exactly the same way as when WinBUGS was used, but the Stata code contains both a function for evaluating the log posterior and an extra function, `pd`, that calculates the predictions using the current state of the chain. In `pd`, new random effects are generated using the current estimate of their standard deviation, and these are combined with the rest of the linear predictor to give the predicted Poisson mean from which random Poisson variables are generated using the standard Stata `rpoisson()` function. In `mcmcrun`, the `predict()` option is used to specify the name of the function that calculates the predictions. These values are added to the output file using names derived from the name of the function; in this case, they are called `pd1`–`pd18`.

begin: model checking in Stata

```
use salm.dta, clear
generate x1 = log(x+10)
generate x2 = x/100
summarize x1
local m1 = r(mean)
replace x1 = x1 - r(mean)
summarize x2
local m2 = r(mean)
replace x2 = x2 - r(mean)

program logpost
   args logp b ipar

   local logtau = log(`b´[1,1])
   tempvar xb
   generate `xb´ = `b´[1,2]+`b´[1,3]*x1+`b´[1,4]*x2
   scalar `logp´ = 0
   forvalues i=1/18 {
      local j = `i´ + 4
      scalar `logp´ = `logp´ + y[`i´]*(`xb´[`i´]+`b´[1,`j´]) ///
           - exp(`xb´[`i´]+`b´[1,`j´]) ///
           - 0.5*`b´[1,`j´]*`b´[1,`j´]*`b´[1,1] + 0.5*`logtau´
   }
```

```
   scalar `logp´ = `logp´ - 0.5*`b´[1,2]*`b´[1,2]/1000000   ///
       - 0.5*`b´[1,3]*`b´[1,3]/1000000 - 0.5*`b´[1,4]*`b´[1,4]/1000000 ///
       - 0.999*`logtau´ - `b´[1,1]/1000
end

program pd, rclass
   args b

   local sd = 1/sqrt(`b´[1,1])
   tempvar xb u
   generate `u´ = rnormal(0,`sd´)
   generate `xb´ = exp(`b´[1,2]+`b´[1,3]*x1+`b´[1,4]*x2+`u´)
   tempname rv
   matrix `rv´ = J(1,18,0)
   forvalues i=1/18 {
      matrix `rv´[1,`i´] = rpoisson(`xb´[`i´])
   }
   return matrix pred = `rv´
end

matrix theta = J(1,22,0)
matrix theta[1,1] = 1
matrix s = J(1,22,0.5)
mcmcrun logpost theta using temp.csv, ///
   samp((mhslogn, sd(s)) 21(mhsnorm, sd(s)))  ///
   burnin(1000) adapt update(10000)  ///
   par(tau beta0 beta1 beta2 u1-u18) savelogp ///
   predict(pd) jpost replace
import delimited temp.csv, clear
replace beta0 = beta0 - `m1´*beta1 - `m2´*beta2
generate sd = 1/sqrt(tau)
save mcmcs.dta, replace
mcmccheck, d(y) p(pd) df(salm.dta) pf(mcmcs.dta) plot(summary)
```

end: model checking in Stata

The function for calculating the log posterior is made reasonably efficient by avoiding any calls to `logdensity`; a further possible improvement is to calculate the conditional distribution dependent on the parameter being updated. However, there are limits to what can be achieved given that the current values of the random effects are stored in a Stata matrix and that the data are stored in Stata variables. Combining them requires loops that are quite slow. Model checking continues in exactly the same way as was used for simulations created by WinBUGS, illustrated by the command for the creation of the summary plot, although the plot is not shown, because it is essentially identical to figure 9.9.

9.6 Residual checking with Mata

The main disadvantage of using Stata rather than WinBUGS to fit the model to the *Salmonella* data is the time that it takes to run; the Stata program takes a couple of minutes on a single-processor desktop computer compared with a few seconds for WinBUGS. This problem can be largely overcome by using Mata. The code below mirrors that used for the Stata program in the last section but calculates the log posterior and predictions using Mata. It is considerably faster than the Stata code and comparable to running WinBUGS. In fact, WinBUGS can run this analysis in just over 3 seconds, while this Mata code takes about 7 seconds. However, the overheads, such as calling WinBUGS and reading the results file, are much less. As was demonstrated in chapter 7, the speed of the Mata program could be improved considerably by more efficient programming, so the balance between WinBUGS and Mata is a fine one; Mata can usually be made to beat WinBUGS for speed, but for most models, less effort is required to create an efficient program in WinBUGS.

begin: model checking in Mata

```
use salm.dta, clear
generate x1 = log(x+10)
generate x2 = x/100
summarize x1
local m1 = r(mean)
replace x1 = x1 - r(mean)
summarize x2
local m2 = r(mean)
replace x2 = x2 - r(mean)
set matastrict off
mata:
mata clear
real scalar logpost(real matrix X,real rowvector t,real scalar p)
{
  sd = 1/sqrt(t[1])
  xb = exp(t[2]:+t[3]*X[,2]+t[4]*X[,3]+t[|5\22|]´)
  logp = logdensity("poisson",X[,1],xb)+
         logdensity("normal",t[|5\22|],0,sd)+
         logdensity("normal",t[|2\4|],0,1000)+
         logdensity("gamma",t[1],0.001,1000)
  return(logp)
}
real rowvector pd(real matrix X,real rowvector t)
{
  sd = 1/sqrt(t[1])
  xb = exp(t[2]:+t[3]*X[,2]+t[4]*X[,3]:+rnormal(18,1,0,sd))
  return(rpoisson(1,1,xb´))
}

end

matrix theta = J(1,22,0)
matrix theta[1,1] = 1
matrix s = J(1,22,0.5)

mcmcrun logpost X theta using temp.csv, ///
  samp((mhslogn, sd(s)) 21(mhsnorm, sd(s))) ///
  burnin(1000) adapt update(10000) jpost dots(0) replace ///
  par(tau beta0 beta1 beta2 u1-u18) savelogp ///
  mata data(X=(y x1 x2) theta s) predict(pd)

import delimited temp.csv, clear
replace beta0 = beta0 - `m1´*beta1 - `m2´*beta2
generate sd = 1/sqrt(tau)
save mcmcm.dta, replace
mcmccheck, d(y) p(pd) dfile(salm.dta) pfile(mcmcm.dta) plot(summary)
```

end: model checking in Mata

9.7 Further reading

Surprisingly, there is not much literature on Bayesian model checking. In a discussion article, Bayarri and Berger (1997) review Bayesian measures of surprise. Gelman et al. (2005) consider the important problem of residual checking in the presence of missing data. Bayarri and Castellanos (2007) consider the checking of hierarchical models.

9.8 Exercises

1. The Stata code

```
set obs 18
egen x = seq(), b(3)
recode x 1=0 2=10 3=33 4=100 5=333 6=1000
generate mu = exp(2 + 0.3*log(x+10) - 0.09*x/100 + rnormal(0,0.25))
generate y = rpoisson(mu)
```

uses a Poisson model with a normal random effect to simulate a set of data that has the same structure as the *Salmonella* data of Margolin, Kaplan, and Zeiger (1981) that was analyzed as a case study in this chapter.

a. Write a Stata program that generates data `x` and `y` as shown above and calls WinBUGS or OpenBUGS to analyze the simulated data using the correct model. Make a summary residual plot. Run your program four times to look at the variation in the plots when the model is correct.

b. Modify the code that simulates the data to introduce an outlier by randomly selecting one of the three readings, y, corresponding to $x = 100$ and replacing it with the value 70. Plot a histogram of the predictive distribution corresponding to the outlier, and note the measures of surprise. Make a summary residual plot. Again run the program four times to see how well the plots detect the outlier.

c. Change the simulation so that there are 10 replicates for each dose and thus 60 observations. Again introduce an outlier of $y = 70$ for one of the readings when $x = 100$, and see how easy it is to detect by repeating the simulation four times and plotting the summary residual plots.

d. Return to a sample size of 18, but choose 18 nonidentical doses ranging between 0 and 1,000, and introduce an outlier of $y = 70$ for the value of x closest to $x = 100$. Repeat the simulation four times, and each time make the summary residual plot to see whether the outlier can be detected.

e. On the basis of these simulations, what advice would you give to an experimenter who is designing a similar dose–response experiment? What advice would you give to a statistician planning to analyze such a dose–response experiment with 18 observations and 6 doses?

2. Case study 8 in chapter 8 fit a nonlinear model to data on lengths and ages of sea cows.

   a. Read the data and keep the same ages, but replace the lengths by randomly generating values that truly follow the nonlinear model using

   ```
   replace y = 2.66 - 0.86^x + rnormal(0,0.1)
   ```

   b. Prepare a Stata program that uses WinBUGS or OpenBUGS to analyze the simulated data by the correct form of nonlinear model together with priors used in chapter 8. Create plots suitable for checking the appropriateness of the model. Save the measures of surprise, and use them together with the original data to create a scatterplot of length against age with the curve of mean predictions superimposed.

   c. Modify the simulation such that the normal errors are replaced by double exponential errors using

   ```
   generate u = -0.1*log(runiform())
   replace u = -u if runiform() < 0.5
   replace y = 2.66 - 0.86^x + u
   ```

   Refit the normal-errors nonlinear model, and see whether you can detect that the error distribution is wrong.

   d. Modify the simulation so that the $U(0.5, 1)$ prior for γ is inappropriate; specifically, let

   ```
   replace y = 2.66 - 0.25^x + rnormal(0,0.1)
   ```

   Refit the normal-errors nonlinear model, and see whether you can detect that the prior excludes the true value of γ.

   e. Modify the simulation so that the relationship is actually linear; specifically, let

   ```
   replace y = 2+0.03*x + rnormal(0,0.1)
   ```

   Refit the normal-errors nonlinear model, and see whether you can detect that the form of the model is wrong.

   f. What conclusions can you draw about the chance of detecting that this nonlinear model is wrong? Did the original analysis use an inappropriate model?

10 Model selection

10.1 Introduction

Real data are not generated from a statistical model, and the most important consequence of this was neatly summarized when Box (1976) wrote an article about the work of his father-in-law, R.A. Fisher. The article contains a lot of sound, general advice about statistics but is best known for expressing the idea that "all models are wrong", later expanded in a technical report to the much-quoted maxim "all models are wrong, but some are useful". In the same article, Box (1976) explained that because all models are wrong, "the scientist cannot obtain a 'correct' one" and that "the scientist must be alert to what is importantly wrong".

Keeping this advice in mind, we should probably expect our selection strategy to do no more than lead us to a model good enough for our needs. Often there will be a range of alternative models, none of which is correct but all of which provide a good fit to our data. In these circumstances, good model-selection strategies should not only lead us to one of the well-fitting models but also provide an indication of the range of plausible alternatives. Indeed, sometimes it is better to think of model selection as an attempt to eliminate models that perform poorly, in order to leave a set of reasonable models that fit the data more or less equally well. This lack of uniqueness causes no problem provided that the conclusions or predictions made under the selected models are all broadly similar; thus an important part of model selection is the investigation of the consistency of any findings across the set of acceptable models.

Naturally enough, Bayesians like to select models based on their probabilities. They might start by using their knowledge of the problem to create a list of plausible models so that they can attach a prior probability to each one. The prior model probabilities are updated in the light of the data, and the model with the highest-posterior probability is identified. Robustness of the conclusions can be checked by seeing whether they would be broadly similar under all the models that have relatively high-posterior probability. The difficult aspects of this approach are the creation of the initial list of plausible models and the allocation of meaningful prior probabilities (remember that we believe that "all models are wrong").

An obvious alternative to selecting one or more of the models from the list is to average some common parameter or prediction across all the models, weighing each contribution by our posterior belief in the model that produced it. This approach, known as model averaging, gives a final answer that incorporates both the precision of each estimate and our uncertainty about the model.

A fundamentally different approach to model selection is to abandon the attempt to specify all models in advance and instead fit a series of models, perhaps using past model fits to guide the choice of the next model. Each time a model is fit, a measure of the goodness of fit is calculated and used to guide our selection. This process should continue until the fit is deemed good enough for our needs or until we can no longer think of plausible new models to try. In some senses, this approach is more natural because it does not require us to assume that one of the models is correct; instead, we can view them all as approximations.

Unfortunately, the sequential approach has its own problems. Because the set of plausible models is never specified, it is more difficult to assess the robustness of the conclusions; perhaps more importantly, it becomes essential to take account of the complexity of the model when making a selection. Starting with any model, one can always generalize it by adding extra parameters; if the original model is nested within the generalization, then the more complex model can never fit worse than the original and so always improves the goodness of fit. Without some penalty for the increased complexity, this would cause us to spiral into increasingly complex models until eventually, it would be possible to make perfect predictions of all the observations. Such overly complex models would add little to our understanding, and the overfitting would probably lead to poor predictions of future observations. Choosing an appropriate penalty for complexity, like choosing the form of the measure of goodness of fit, is really a matter of judgment. Our hope is that sensible selection criteria leads to similar model choices. Box (1976) even had something to say on overly elaborate models pointing out that "[j]ust as the ability to devise simple but evocative models is the signature of the great scientist[,] so overelaboration and overparameterization is often the mark of mediocrity".

10.2 Case study 11: Choosing a genetic model

Kirke et al. (2004) reported an Irish case–control study of 395 patients with neural tube defects (NTD) and 848 healthy controls. All the subjects had their genotype measured at a particular location in the *MTHFR* gene, where the genome can take one of two forms, usually referred to as C and T. Because everyone inherits two copies of every gene, one from each parent, the subjects will have the CC, CT, or TT genotype. If this variant of the *MTHFR* gene is associated with NTD, then the proportions of people with each genotype will vary between cases and controls. In particular, if T is the harmful version of the variant, then TTs should be relatively more common in the cases. The actual data from the study are shown in table 10.1.

Table 10.1. Number of people with each of the three genotypes

	CC	CT	TT
Cases	151	171	73
Controls	439	326	83

10.2.1 Plausible models

Experience with other genetic variants and other diseases suggests that the pattern of increased risk in people with the intermediate or heterozygous genotype (CT) often follows one of three patterns. For some genetic diseases, heterozygotes are at the same risk as the TTs. In this case, the T allele is referred to as dominant, while for other diseases, heterozygotes are not at increased risk at all, and the T allele is referred to as recessive. Yet for other diseases, the risk to heterozygotes is half (on a log scale) of that experienced by the high-risk homozygotes. This is often referred to as the (log) additive model. Other patterns are found but far less frequently.

Representing the genotype proportions in the population of all potential controls by p_1, p_2, and p_3 and the proportions in all potential cases by q_1, q_2, and q_3, these models can be expressed in terms of the two odds ratios

$$\mathrm{OR}_{21} = \frac{q_2 p_1}{p_2 q_1} \qquad \mathrm{OR}_{31} = \frac{q_3 p_1}{p_3 q_1}$$

An initial assessment of these models could be obtained by calculating the corresponding odds ratios using the data in table 10.1. These are $(171 \times 439)/(326 \times 151) = 1.52$ and $(73 \times 439)/(83 \times 151) = 2.56$, so the additive model appears to be a good option, although this calculation does not give any indication of the strength of the evidence for our choice.

If we include the possibility that the genetic variant is not associated with NTD, then there are four plausible models:

- M_1 No effect: $\mathrm{OR}_{21} = 1$, $\mathrm{OR}_{31} = 1$
- M_2 Recessive effect: $\mathrm{OR}_{21} = 1$, $\mathrm{OR}_{31} \neq 1$
- M_3 Additive effect: $\mathrm{OR}_{21} = \sqrt{\mathrm{OR}_{31}}$, $\mathrm{OR}_{31} \neq 1$
- M_4 Dominant effect: $\mathrm{OR}_{21} = \mathrm{OR}_{31}$, $\mathrm{OR}_{31} \neq 1$

Our problem is to decide between these four competing models and to quantify our confidence in the final choice. Because we are conducting a Bayesian analysis, the selection will be based on our posterior beliefs in the four models, and those posterior beliefs will depend on our prior beliefs and the evidence from the data.

Suppose that our prior beliefs in the models are denoted by $p(M_j)$ $j = 1, \ldots, 4$, and our prior beliefs about the parameters of those models by $p(\theta|M_j)$. Basic probability calculations give the marginal likelihoods for each model,

$$p(y|M_j) = \int p(y|\theta, M_j)p(\theta|M_j)d\theta$$

and the posterior model probabilities are given by

$$p(M_j|y) \propto p(y|M_j)P(M_j) = p(M_j) \int p(y|\theta, M_j)p(\theta|M_j)d\theta$$

where the constant of proportionality is obtained using the fact that the four posterior model probabilities must sum to 1.

The simplicity of this approach must not lead us to suppose that these calculations are easy to perform, because in general, they are not. Several methods have been proposed for integrating over the full set of parameters, but as we will see, none of them is entirely satisfactory.

10.2.2 Bayes factors

To compare two of the models, we might choose to calculate the ratio of their posterior probabilities:

$$\begin{aligned}\frac{p(M_2|y)}{p(M_1|y)} &= \frac{\int p(y|\theta, M_2)p(\theta|M_2)d\theta}{\int p(y|\theta, M_1)p(\theta|M_1)d\theta} \frac{p(M_2)}{p(M_1)} \\ &= \mathrm{BF}_{21} \frac{p(M_2)}{p(M_1)}\end{aligned}$$

This equation shows how our comparative belief in those two models has changed because of the data. The factor that converts the prior probability ratio into the posterior probability ratio is known as the Bayes factor (BF).

$$\mathrm{BF}_{21} = \frac{\int p(y|\theta, M_2)p(\theta|M_2)d\theta}{\int p(y|\theta, M_1)p(\theta|M_1)d\theta}$$

A BF is the ratio of two marginal likelihoods and is close in spirit to the likelihood ratio of traditional statistics in that it tells us how strongly the data favor one model over the other. It is independent of our prior beliefs about the models but not of our prior beliefs about the parameters in those models.

An attractive property of the BF is that once it is known, it can be combined with our prior model probabilities to lead directly to the posterior probabilities of the models. Consider the case of two competing models:

$$p(M_1|y) = \frac{p(y|M_1)p(M_1)}{p(y|M_1)p(M_1) + p(y|M_2)p(M_2)} = \frac{p(M_1)}{p(M_1) + \mathrm{BF}_{21}p(M_2)}$$

The posterior model probabilities have the advantage of being on a well-understood (0,1) scale, but the scale of the BF is much less intuitive. To help with the interpretation of BFs, several authors have given guidelines of the type displayed in table 10.2. When there are two models and we have no prior preference for either, then a BF of 20 implies that the posterior probability of the favored model is $20/(1+20)\approx 0.95$. So even when BF $=20$, if we have equal priors, we acknowledge a 5% chance that the less favored model might actually be correct. For many people, this is the point at which they begin to be persuaded to use one model over the other, although BFs closer to 100 are needed for really convincing evidence.

Table 10.2. Guide to interpreting BFs; the posterior model probabilities $p(M_2|y)$ assume that there was no preference for either model prior to observing the data

BF_{21}	BF_{12}	$p(M_2\|y)$	Strength of evidence
0.005	200	0.005	very strong
0.05	20	0.05	quite strong
0.33	3	0.25	weak
1	1	0.5	negligible
3	0.33	0.75	weak
20	0.05	0.95	quite strong
200	0.005	0.995	very strong

10.3 Calculating a BF

The calculation of a BF involves integration over the model parameters, and because the BF is a ratio, those integrals need to be evaluated accurately. The heart of the calculation is to find the marginal likelihood for each of the different models:

$$\int p(y|\theta, M_j)p(\theta|M_j)d\theta$$

One obvious approach is to generate a sample of values $\theta^{(i)}$, where $i=1,\ldots,m$, from the prior $p(\theta|M_j)$ and evaluate $p(y|\theta^{(i)}, M_j)$ at each point so that the integral can be approximated by

$$\int p(y|\theta, M_j)p(\theta|M_j)d\theta \approx \frac{1}{m}\sum_{i=1}^{m} p(y|\theta^{(i)}, M_j)$$

Unfortunately, unless the prior on θ is reasonably informative, it is likely that $p(y|\theta^{(i)}, M_j)$ will be extremely small for most of the sampled parameter values. In this case, only a tiny proportion of those values will contribute to the average; thus this method can be inefficient and require large samples.

When the priors are vague, a more realistic approach is to use importance sampling, whereby the sample of parameter values, $\theta^{(i)}$, is drawn from an arbitrary distribution $p'(\theta)$, and the integral is approximated by the weighted average,

$$\int p(y|\theta, M_j)p(\theta|M_j)d\theta \approx \frac{\sum_{i=1}^{m} w_i p(y|\theta^{(i)}, M_j)}{\sum_{i=1}^{m} w_i}$$

where $w_i = p(\theta^{(i)}|M_j)/p'(\theta^{(i)})$. An obvious choice for $p'(\theta)$ is the posterior distribution of θ or an approximation to it: this concentrates the points at places where $p(y|\theta^{(i)}, M_j)$ is relatively large. Simple algebraic substitution shows that when points are drawn from the posterior, the importance sampling formula reduces to the harmonic mean:

$$\int p(y|\theta, M_j)p(\theta|M_j)d\theta \approx \left\{ \frac{1}{m} \sum_{i=1}^{m} \frac{1}{p(y|\theta^{(i)}, M_j)} \right\}^{-1}$$

This approximate method often works well but can be unstable when some of the $p(y|\theta^{(i)}, M_j)$ are small.

A third approach to calculating the BFs is to set up a Markov chain Monte Carlo (MCMC) algorithm to calculate the posterior model probabilities, $p(M_j|y)$, directly. Working backward, one can obtain the BF as the change in probability ratios:

$$\text{BF}_{21} = \frac{p(M_2|y)/p(M_1|y)}{p(M_2)/p(M_1)}$$

This approach is ideally suited to MCMC analysis because the chosen model can be thought of as just another parameter, M. If the currently chosen model is $M = j$ and the other parameter values are denoted by θ, then we might propose a move to model $M = k$ and new parameters θ' using a proposal distribution $g()$, accepting the move in a Metropolis–Hastings step with acceptance probability:

$$\frac{p(y|M = k, \theta')p(M = k, \theta')g(M = k, \theta', M = j, \theta)}{p(y|M = j, \theta)p(M = j, \theta)g(M = j, \theta, M = k, \theta')}$$

It is important to clarify exactly what is meant by θ in this expression. It is the vector of parameters for all the models. Perhaps some parameters are relevant only when $M = j$ and others apply only when $M = k$, yet all the parameters must be updated at every iteration.

Often, it is convenient to factorize the priors, so we first specify M and then specify the model parameters conditional on that model choice; in this case, the Metropolis–Hastings acceptance probability becomes

$$\frac{p(y|M = k, \theta')p(\theta'|M = k)p(M = k)g(M = k, \theta', M = j, \theta)}{p(y|M = j, \theta)p(\theta|M = j)p(M = j)\, g(M = j, \theta, M = k, \theta')}$$

It is now clear that we must specify our prior beliefs about all parameters, including those that apply to model M_j, when we believe that the "correct" model is M_k. These artificial priors are usually referred to as pseudopriors, and as with a proposal distribution in a Metropolis–Hastings algorithm, they will not influence the results but they can have a big impact on the chain's efficiency. The chain will converge more quickly if the pseudoprior for the parameters of model M_j is chosen to be similar to the posterior distribution of those parameters that applies when M_j is the correct model.

The idea of pseudopriors was first proposed by Carlin and Chib (1995) in an algorithm that is a slight variation on the Metropolis–Hastings version described above. Instead of using the acceptance probability, they suggested calculating $A_j = p(y|M = j, \theta)p(\theta|M = j)p(M = j)$ for each model so that

$$p(M = k|\theta, y) = \frac{A_k}{\sum A_j}$$

The next model in the chain can then be chosen at random with these probabilities.

10.4 Calculating the BFs for the NTD case study

To calculate posterior model probabilities and BFs for the NTD study, we must first specify our model for the data and describe our prior beliefs. Assuming that the subjects are independent, then a multinomial likelihood is

$$p(y|p_j, q_j) \propto p_1^{439} p_2^{326} p_3^{83} q_1^{151} q_2^{171} q_3^{73}$$

and a convenient prior for the probabilities in controls is provided by the Dirichlet distribution,

$$p(p_1, p_2, p_3) = \frac{\Gamma(\alpha_1 + \alpha_2 + \alpha_3)}{\Gamma(\alpha_1)\Gamma(\alpha_2)\Gamma(\alpha_3)} p_1^{\alpha_1 - 1} p_2^{\alpha_2 - 1} p_3^{\alpha_3 - 1}$$

where α_1, α_2, and α_3 are chosen as if they were the numbers with each genotype in an imaginary study. This genetic variant of the *MTHFR* gene has been studied many times in conjunction with other diseases, and comparable control populations show genotype proportions in the approximate ratio 5:4:1. If we judge that this prior information is equivalent to a real study of about 50 subjects, then we might set $\alpha_1 = 25$, $\alpha_2 = 20$, and $\alpha_3 = 5$.

Genetic effects tend to be quite small: an odds ratio comparing TT with CC might reach 2 or 3 but quite possibly would be smaller than that. If we write $\theta_2 = \log \text{OR}_{21}$ and $\theta_3 = \log \text{OR}_{31}$, a reasonable choice of a prior under models M2, M3, and M4 might be

$$p(\theta_3) \sim N(0.7, 0.5)$$

from which the prior on θ_2 is implicit but varies according to the model under consideration. We could have chosen a different prior on θ_3 under each model, but in the absence of more detailed knowledge, we will adopt the common prior.

Once these odds ratios have been specified, the genotype probabilities in cases can be obtained using the following relationships derived from the definitions of the odds ratios:

$$q_1 = \frac{p_1}{p_1 + p_2 e^{\theta_2} + p_3 e^{\theta_3}} \quad q_2 = \frac{p_2 e^{\theta_2}}{p_1 + p_2 e^{\theta_2} + p_3 e^{\theta_3}} \quad q_3 = \frac{p_3 e^{\theta_3}}{p_1 + p_2 e^{\theta_2} + p_3 e^{\theta_3}}$$

Thus the likelihood is completely determined by the control proportions, p_j, and the log odds-ratio, θ_3.

For this particular problem, the method of integration based on samples from the prior is perfectly practical because we have chosen to use realistic priors that are not needlessly vague. The Stata code below calculates the log of the marginal likelihood, $\log\{P(y|M_j)\}$, for each of the four models. Samples from the Dirichlet distribution are obtained by using the Stata function for generating random numbers from a gamma distribution and then normalizing the resulting random numbers to sum to 1.

begin: BFs in Stata

```
clear
set obs 500000
set seed 546109
generate p1 = rgamma(25,1)
generate p2 = rgamma(20,1)
generate p3 = rgamma(5,1)
generate sp = p1+p2+p3
replace p1 = p1/sp
replace p2 = p2/sp
replace p3 = p3/sp
generate theta = rnormal(0.7,0.5)

generate logp = 439*log(p1)+326*log(p2)+83*log(p3)
generate OR3 = exp(theta)
* Model M1
generate logp1 = logp + 151*log(p1)+171*log(p2)+73*log(p3)

* Model M2
generate OR2 = 1
generate sq = p1+OR2*p2+OR3*p3
generate q1 = p1/sq
generate q2 = OR2*p2/sq
generate q3 = OR3*p3/sq
generate logp2 = logp + 151*log(q1)+171*log(q2)+73*log(q3)

* Model M2
replace OR2 = exp(theta/2)
replace sq = p1+OR2*p2+OR3*p3
replace q1 = p1/sq
replace q2 = OR2*p2/sq
replace q3 = OR3*p3/sq
generate logp3 = logp + 151*log(q1)+171*log(q2)+73*log(q3)

* Model M3
replace OR2 = OR3
replace sq = p1+OR2*p2+OR3*p3
replace q1 = p1/sq
replace q2 = OR2*p2/sq
```

```
replace q3 = OR3*p3/sq
generate logp4 = logp + 151*log(q1)+171*log(q2)+73*log(q3)

* Baseline for exponentiation
summarize logp3
local m = r(max)
* Marginal Likelihoods
forvalues i=1/4 {
   generate m`i´ = exp(logp`i´-`m´)
   summarize m`i´
   local m`i´ = r(mean)
}
display "BF31" %10.0f `m3´/`m1´
display "BF32" %10.0f `m3´/`m2´
display "BF34" %10.0f `m3´/`m4´
```

end: BFs in Stata

This program calculates and stores $\log\{P(y|\theta^{(i)}, M_j)\}$ at each sampled point, while the integral approximation requires

$$\frac{1}{m}\sum_{j=1}^{m} p(y|\theta^{(i)}, M_j)$$

The values of $\log\{P(y|\theta^{(i)}, M_j)\}$ given by the program are of the order of $-1{,}240$, so direct exponentiation loses all precision. However, because BFs only require ratios of the marginal likelihoods, it is sufficient to average $\exp[\log\{P(y|\theta^{(i)}, M_j)\} - C]$, where C is an arbitrary constant that reflects typical sizes of the values, here taken to be the maximum of $\log\{P(y|\theta^{(i)}, M_3)\}$.

Because model 3 is the most likely to have generated these data, it is convenient to give the results for that model relative to the others. Rounding the calculated values produced $\text{BF}_{31} = 322,000$, $\text{BF}_{32} = 130$, and $\text{BF}_{34} = 62$. There is not much difficulty in excluding M_1, the model that supposed that the variant was unrelated to NTD in favor of model 3. The evidence in the data certainly points to an additive effect, although based on the data, a dominant effect (model 4) is still just a possibility. In practice, the model that would be selected depends not just on these BFs but also on what we thought prior to collecting the data. If we had thought that all four models were equally likely, then

$$P(M_3|y) = \frac{1}{1 + 1/330000 + 1/130 + 1/65} = 0.98$$

We would be 98% sure that the additive model is "correct".

Alternatively, WinBUGS could be used to generate the sample from the prior using the code

```
model {
   p[1:3] ~ ddirich(a[])
   theta  ~ dnorm(0.7,4)
}
```

with the data file containing

```
list(a=c(25,20,5))
```

and the initial values file containing

```
list(p=c(0.5,0.4,0.1), theta=0.7)
```

OpenBUGS denotes the Dirichlet distribution by `ddirich()`, while WinBUGS calls it `ddirch()`, so this code needs to be changed to work in WinBUGS. Stata code such as

begin: WinBUGS priors

```
wbsscript using script.txt, ///
   model(priormodel.txt) data(data.txt) init(init.txt) ///
   burn(0) update(50000) thin(5) set(p theta) coda(BF) ///
   replace openbugs
wbsrun using script.txt, openbugs
wbscoda using BF, clear openbugs
```

end: WinBUGS priors

runs the model and generates the necessary sample from the prior, after which the calculation progresses as before. There is no advantage in using WinBUGS for this problem, but it does have a wider range of random-number generators than Stata and may be useful for some other models.

The method of sampling from the posterior and calculating the harmonic mean is a useful technique for larger problems or problems with vaguer priors, although this approach does tend to be heavily influenced by a few extreme points. For the NTD data, this method can be implemented as shown below. The first four sections show the model files for fitting the different models. This is followed by the Stata code needed to call WinBUGS and then process the resulting MCMC chains.

begin: BF by harmonic mean

```
M1: MODEL FILE modelBF1.txt
model {
    p[1:3] ~ ddirich(a[])
    theta <- 0
    for (i in 1:3) {
       q[i] <- p[i]
    }
    x[1:3] ~ dmulti(p[1:3],Nx)
    y[1:3] ~ dmulti(p[1:3],Ny)
}
M2: MODEL FILE modelBF2.txt
model {
    p[1:3] ~ ddirich(a[])
    theta ~ dnorm(0.7,4)
    OR3 <- exp(theta)
    OR2 <- 1
    s <- p[1] + OR2*p[2] + OR3*p[3]
    q[1] <- p[1]/s
    q[2] <- OR2*p[2]/s
    q[3] <- OR3*p[3]/s
    x[1:3] ~ dmulti(p[],Nx)
    y[1:3] ~ dmulti(q[],Ny)
}
M3: MODEL FILE modelBF3.txt
model {
    p[1:3] ~ ddirich(a[])
    theta ~ dnorm(0.7,4)
    OR3 <- exp(theta)
    OR2 <- exp(theta/2)
    s <- p[1] + OR2*p[2] + OR3*p[3]
    q[1] <- p[1]/s
    q[2] <- OR2*p[2]/s
    q[3] <- OR3*p[3]/s
    x[1:3] ~ dmulti(p[],Nx)
    y[1:3] ~ dmulti(q[],Ny)
}
M4: MODEL FILE modelBF4.txt
model {
    p[1:3] ~ ddirich(a[])
    theta ~ dnorm(0.7,4)
    OR3 <- exp(theta)
    OR2 <- OR3
    s <- p[1] + OR2*p[2] + OR3*p[3]
    q[1] <- p[1]/s
    q[2] <- OR2*p[2]/s
    q[3] <- OR3*p[3]/s
    x[1:3] ~ dmulti(p[],Nx)
    y[1:3] ~ dmulti(q[],Ny)
}
clear
set more off
wbslist (a=c(25,20,5), x=c(439,326,83), Nx=848, y=c(151,171,73), Ny=395) ///
   using dataBF.txt, replace
wbslist (p=c(0.5,0.4,0.1), theta = 0.7) using initBF.txt, replace
forvalues M=1/4 {
   wbsscript using script.txt,  ///
     model(modelBF`M'.txt) data(dataBF.txt) init(initBF.txt)  ///
     burn(10000) update(50000) set(p q) coda(BF) replace openbugs
```

```
    wbsrun using script.txt, openbugs time
    wbscoda using BF, clear openbugs
    generate logp`M´ = 439*log(p_1)+326*log(p_2)+83*log(p_3) + ///
                  151*log(q_1)+171*log(q_2)+73*log(q_3)
    keep logp`M´
    if `M´ > 1 quietly merge 1:1 _n using BFH.dta, nogen
    save BFH.dta, replace
}

use BFH.dta, clear
summarize logp3
local m = r(max)
generate m1 = 1/exp(logp1-`m´)
generate m2 = 1/exp(logp2-`m´)
generate m3 = 1/exp(logp3-`m´)
generate m4 = 1/exp(logp4-`m´)
summarize m1
local m1 = 1/r(mean)
summarize m2
local m2 = 1/r(mean)
summarize m3
local m3 = 1/r(mean)
summarize m4
local m4 = 1/r(mean)
display %10.0f `m3´/`m1´ %10.0f `m3´/`m2´ %10.0f `m3´/`m4´
```

end: BF by harmonic mean

A single run of this program gave BFs equal to 386,000, 105, and 35, but switching to `seed(2)` gave 510,000, 161, and 52. Clearly, the harmonic mean is not very stable, and it is difficult to get a reliable value for the BF using this method.

Finally, the WinBUGS program given below runs the Carlin and Chib pseudoprior algorithm for the NTD data. This method fits all four models at once and allows the chain to switch between models. If we were to start with equal priors on all four models, then the chain would almost never visit model 1. This prevents us from calculating BFs involving model 1, so to force a more even spread of visits to the four models, we placed artificial prior probabilities in the ratio 0.9979:0.001:0.0001:0.001 on the four models. As a result, the 500,000 simulations spend about 10% of the time in model 1 and about 30% of the time in each of the other models. This enables us to obtain more stable estimates of the full set of BFs, which are, of course, measures of change in belief caused by the data and so are independent of the choices of prior model probabilities. A single run of the program gave BFs for model 3 relative to the other models of 297,000, 133, and 64.

begin: pseudopriors model file

```
model {
    M ~ dcat(alpha[1:4])
    p[1:3] ~ ddirich(a[1:3])

    m2[1] <- 0.75 m2[2] <- 0.7  m2[3] <- 0.75 m2[4] <- 0.75
    t2[1] <- 25   t2[2] <- 4    t2[3] <- 25   t2[4] <- 25

    m3[1] <- 0.9  m3[2] <- 0.9  m3[3] <- 0.7  m3[4] <- 0.9
    t3[1] <- 25   t3[2] <- 25   t3[3] <- 4    t3[4] <- 25

    m4[1] <- 0.6  m4[2] <- 0.6  m4[3] <- 0.6  m4[4] <- 0.7
    t4[1] <- 25   t4[2] <- 25   t4[3] <- 25   t4[4] <- 4

    theta[1,1] <- 0
    theta[1,2] <- 0

    theta[2,1] <- 0
    theta[2,2] ~ dnorm(m2[M],t2[M])

    theta[3,2] ~ dnorm(m3[M],t3[M])
    theta[3,1] <- theta[3,2]/2

    theta[4,2] ~ dnorm(m4[M],t4[M])
    theta[4,1] <- theta[4,2]


    OR2 <- exp(theta[M,1])
    OR3 <- exp(theta[M,2])

    s <- p[1] + OR2*p[2] + OR3*p[3]
    q[1] <- p[1]/s
    q[2] <- OR2*p[2]/s
    q[3] <- OR3*p[3]/s

    x[1:3] ~ dmulti(p[1:3],Nx)
    y[1:3] ~ dmulti(q[1:3],Ny)
}
```

end: pseudopriors model file

Before considering the details of this WinBUGS code, notice that it is possible to specify multiple assignments on the same line without the need for a special separator. The code creates a 4×2 matrix of log odds-ratios called *theta*. This contains logOR2 and logOR3 under the four models.

For illustration, consider the updating of the parameters of model 2. As the effect is recessive, the first log odds-ratio is always zero, and the second log odds-ratio is updated using a normal prior with parameters `m2[M]` and `t2[M]`. When the current model is `M` $=$ 2, this must correspond to our true prior; that is, `m2[2]` $= 0.7$ and `t2[2]` $= 4$. When `M` $=3$, we still need to update the value of `theta[2,2]`; thus we must give it a pseudoprior. By fitting model 2 on its own, we can discover that the values of the log odds-ratio have a posterior centered around 0.9 with a precision of about 25, so that distribution is used for the pseudoprior. When `M` $=2$, `theta[2,2]` will be updated by combining the true prior with the data; otherwise, `theta[2,2]` will be updated from the pseudoprior alone. This means that even when we are not in model 2, we continue to give theta[2,2] realistic values so that the algorithm is always open to the possibility of switching to M = 2, hence ensuring good mixing.

The data file to accompany the pseudopriors code should contain

```
list(alpha=c(0.9979,0.001,0.0001,0.001), x=c(439,326,83),
   y=c(151,171,73), a=c(25,20,5), Nx=848, Ny=395)
```

The Stata code needed to call WinBUGS is

begin: pseudopriors Stata code

```
wbslist (alpha=c(0.9979,0.001,0.0001,0.001), x=c(439,326,83), ///
    y=c(151,171,73), a=c(25,20,5), Nx=848, Ny=395) using dataCB.txt, replace
matrix input theta = (. . \ . 0.7 \ . 0.7 \ . 0.7)
matrix input p = (0.4, 0.4, 0.2)
wbslist(matrix theta p) (M=1) using initCB.txt, replace
wbsscript using script.txt,                              ///
   model(modelCB.txt) data(dataCB.txt) init(initCB.txt) ///
   burn(0) update(100000) thin(5) log(logCB.txt)        ///
   set(p theta M) coda(CB) openbugs replace
wbsrun using script.txt, openbugs
type logCB.txt
wbscoda using CB, clear openbugs
forvalues i=1/4 {
   count if M == `i´
   local n`i´ = r(N)
}
display (`n3´/`n1´)/(0.0001/0.9979)
display (`n3´/`n2´)/(0.0001/0.001)
display (`n3´/`n4´)/(0.0001/0.001)
```

end: pseudopriors Stata code

In this code, we give initial values for the matrix theta. Because theta is a mixture of assigned values and simulated values, we must be careful to place missing values in the places corresponding to the assignments. Because theta is a matrix, the values need to be specified as a two-way structure. As always, the convergence of the chain needs checking, although in this case, the initial values are so good that there is no early drift, and a burn-in is not really needed. Because we based our pseudopriors on the known posterior distributions of the separate models, the trace of theta[2,2] looks like one complete sequence, but of course, it is actually made up of a sequence when M = 2 and a separate sequence when M $\neq$ 2. For the purpose of assessing convergence, we need to concentrate on the chain for theta[2,2] when M = 2 because the remainder will merely consist of samples from the pseudoprior. Once the program has run, the BFs are estimated as the changes between the prior odds and the posterior odds.

The three methods demonstrate that although it is easy to approximate BFs, it is not at all easy to get precise values. However, because our interpretation of BFs is unaffected by small differences, an approximate value is usually sufficient.

10.5 Robustness of the BF

Often the priors for the model parameters are chosen rather arbitrarily in the hope that they will be noninformative, or perhaps they only represent crude approximations to the beliefs of an individual or of a group of people. In such circumstances, it would not be sensible to treat the priors as being known precisely. The probabilities, $P(y|M)$, are obtained by integrating over those prior distributions, so one must confirm that small changes to the priors on the model parameters do not result in noticeable differences to the BFs.

To investigate the robustness of the BFs of the NTD study, we performed a sensitivity analysis in which the priors were varied slightly and the BFs were recalculated. In table 10.3, the estimated BFs for the four genetic models are given assuming that the Dirichlet prior on the three control proportions remains the same but the prior on the log odds-ratio, θ_3, varies slightly. The estimated BFs were obtained by random sampling from the prior and so are themselves subject to some sampling error. However, sufficient points were taken so that the error in estimation is only of the order of 1%–2%, not nearly enough to explain the observed differences in the BFs.

Table 10.3. Robustness of the BFs to changes in the prior on the logOR

Prior p(θ_3)	BF$_{31}$	BF$_{32}$	BF$_{34}$
$N(0.7, 0.5)$	323,000	131	62
$N(0.6, 0.5)$	280,000	119	53
$N(0.8, 0.5)$	347,000	136	72
$N(0.7, 0.25)$	485,000	110	57
$N(0.7, 1.0)$	180,000	138	65

The BFs in table 10.3 all lead us to the same model choice, but the exact values do vary because the prior changes, and even greater variability would have been obtained had the prior on the control proportions also been varied. The lesson is that we should treat BFs as broad indicators of the evidence in the data and not as precise measures, partly because BFs are difficult to calculate accurately and partly because they are sensitive to the choice of prior for the model parameters.

10.6 Model averaging

As a by-product of running the Carlin and Chib algorithm, we obtain parameter estimates under each of the different models. If the program is run with equal priors on each of the four models, then the algorithm spends almost all the time in the additive model, and the average result across all models is almost identical to the result for model 3. Because this case is rather uninstructive, we illustrate the impact of model averaging by supposing that prior to the analysis, we favored a dominant genetic model, allocating it a prior probability of 85% and giving only 5% probability to each of the other three models. The program was rerun with these priors, and the estimated value of the log odds-ratio θ_3 was obtained. Under the dominant model, the estimate of θ_3 is a compromise between the log odds-ratios for the CT genotype and the TT genotype. This makes it somewhat smaller than the estimate under the recessive or additive models as the results below show.

```
. mcmcstats logor
----------------------------------------------------------------------------
Parameter           n      mean        sd       sem    median         95% CrI
----------------------------------------------------------------------------
 logor         500000     0.818     0.208    0.0008     0.833 (  0.405,   1.193 )
----------------------------------------------------------------------------

. mcmcstats logor if M == 2
----------------------------------------------------------------------------
Parameter           n      mean        sd       sem    median         95% CrI
----------------------------------------------------------------------------
 logor           3090     0.727     0.167    0.0051     0.724 (  0.401,   1.045 )
----------------------------------------------------------------------------

. mcmcstats logor if M == 3
----------------------------------------------------------------------------
Parameter           n      mean        sd       sem    median         95% CrI
----------------------------------------------------------------------------
 logor         389887     0.890     0.165    0.0007     0.890 (  0.567,   1.211 )
----------------------------------------------------------------------------

. mcmcstats logor if M == 4
----------------------------------------------------------------------------
Parameter           n      mean        sd       sem    median         95% CrI
----------------------------------------------------------------------------
 logor         107016     0.556     0.121    0.0008     0.557 (  0.321,   0.794 )
----------------------------------------------------------------------------
```

The additive model is still preferred, despite our prior belief in the dominant model. The posterior probability of the additive model is 389887/500000 or about 80%; thus if we adopt this model, our posterior estimate of the log odds-ratio would be 0.890. However, because of our prior belief, the data are not strong enough to rule out the possibility of a dominant model under which we would have estimated that the log odds-ratio is 0.556. Rather than choose a single model, we could base the estimate on the full sample, effectively averaging across the models in proportion to their posterior probability. Such model averaging produces an estimate of 0.818 with a wider credible interval reflecting our uncertainty about the genetic model.

Figure 10.1 shows the posterior distributions for the log odds-ratio of TT versus CC under model averaging (solid line) and under the additive and dominant models (dashed lines). Under the separate genetic models, the densities have been scaled to have an area equal to their posterior probability.

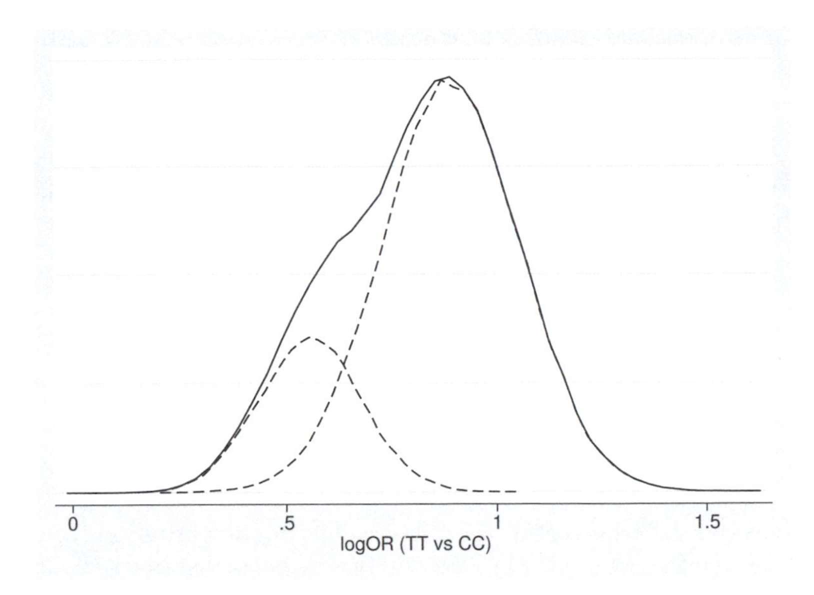

Figure 10.1. Log odds-ratio estimate averaged across the four models

10.7 Information criteria

An alternative to comparing a predefined set of models with BFs is to calculate a measure of goodness of fit each time a new model is tried. Selection can then be based on a comparison of those measures. Of course, a complex model with many parameters would be expected to fit better than a simpler model with few parameters, even when the simpler model fits perfectly well. It is important, therefore, to make some adjustment for the complexity of each model. For basic models, the complexity is often adequately captured by the number of parameters, but when the model is hierarchical and there are many unobserved effects that are constrained to come from a common distribution, it can be very difficult to decide how many effective parameters there are.

The most commonly used non-Bayesian measure of fit is the Akaike information criterion (AIC), which is calculated as minus the maximum of the log likelihood of the observed data penalized by twice the number of parameters in the model, K.

$$\text{AIC} = -2\log(P(y|\widehat{\theta}) + 2K$$

The penalty $2K$ can be thought of either as a measure of the model's complexity or as an adjustment for a slightly over optimistic assessment of fit produced when the same data are used both to estimate the parameters and to measure that fit. The model with the lowest AIC is preferred, but generally, there is no way of knowing what difference in AIC might just have arisen by chance. This has led to rather vague guidelines, such as not relying on differences smaller than 4 or 5.

A different penalty was suggested by Schwarz (1978), and this produces the Bayesian information criterion (BIC),

$$\text{BIC} = -2\log(P(y|\widehat{\theta}) + K\log(n)$$

where n is the number of observations. Despite the name, this measure also depends on the maximum of the log likelihood and is essentially non-Bayesian.

The main advantage of the BIC is that assuming that the impact of the priors is not too great, differences in BIC can be related back to the BF using the approximation

$$\text{BF}_{21} \approx \exp((\text{BIC}_2 - \text{BIC}_1)/2)$$

This gives us an immediate scale for interpretation. A difference in BIC between two models of 6 is roughly equivalent to a BF of 20, and a difference in BIC of 10 is approximately equivalent to a BF of 150.

A fully Bayesian criterion, called the deviance information criterion (DIC), was suggested by Spiegelhalter et al. (2002). This criterion does not have a particularly strong theoretical basis, and its use for complex models with many random effects has been questioned. Nevertheless, it is simple to calculate and is provided as an option within WinBUGS, so it has become very popular.

As with the other information criteria, Spiegelhalter et al. (2002) base the DIC on the deviance, $D(\theta) = -2\log\{P(y|\theta)\}$, although here it is viewed as a function of the parameters. Non-Bayesian criteria evaluate the deviance at its maximum, but for a Bayesian, it is more natural to obtain a measure of fit, $\overline{D}$, by integrating the deviance over our posterior beliefs about the parameters:

$$\overline{D} = \int D(\theta)P(\theta|y)d\theta$$

This quantity is easily approximated as part of an MCMC algorithm by calculating the deviance for each simulated set of parameter values, $\theta^{(i)}$, and then averaging them.

$$\overline{D} \approx \frac{1}{m}\sum_{i=1}^{m} D(\theta^{(i)})$$

As a measure of model complexity, Spiegelhalter et al. (2002) suggest

$$p_D = \overline{D} - D(\overline{\theta})$$

where $\overline{\theta}$ represents the posterior mean estimates of θ. The full deviance information criterion can now be calculated as a penalized measure of fit:

$$\text{DIC} = \overline{D} + p_D$$

Parallels with the AIC and BIC are clearly very strong. Again we seek a low DIC, but deciding when a change in DIC can be ignored is just as difficult as it was for the other measures of information, and people tend to use similar vague guidelines, such as selecting the minimum DIC but not putting too much reliance on differences below 5.

One of the strengths of p_D as a measure of complexity is that it provides a solution to the problem of measuring the effective number of parameters in complex hierarchical models in which there are unobserved effects from a common distribution. Limited experience with p_D suggests that it provides a good solution to this problem provided that the total number of parameters, including the unobserved or random effects, is small compared with the number of observations. Otherwise, p_D may underpenalize the model fit.

10.8 DIC for the genetic models

The DIC is calculated by WinBUGS, so it is easy to obtain the measures for each of the four models by fitting each one in turn. The values of the DIC will be written to the log file if the `dic` option is added to `wbsscript`.

The results of running the four models are shown in table 10.4. Because there are no random effects to complicate the calculations, the values of p_D are close to the number of parameters that have to be estimated. Comparison of the DICs confirms a strong preference for model 3 with no support for model 1.

Table 10.4. DIC for the NTD models

Model	p_D	DIC
Model 1: No effect	1.9	56.9
Model 2: Recessive	2.8	41.0
Model 3: Additive	2.8	31.2
Model 4: Dominant	2.8	38.7

10.9 Starting points

Kass and Raftery (1995) give a thorough review of BFs. Many authors have written nonmathematical introductions to BFs, of which Goodman (1999) is a good example. Different methods for computing BFs are considered by Han and Carlin (2001), and the related problem of estimating posterior model probabilities using MCMC is discussed in Dellaportas, Forster, and Ntzoufras (2002). Hoeting et al. (1999) present a tutorial on Bayesian model averaging, and that article is accompanied by a interesting discussion.

Spiegelhalter et al. (2002) introduced the DIC, but perhaps the best paper for understanding when and why it works is Plummer (2008).

10.10 Exercises

1. Case study 3 looked at polyp counts in women recruited into a cancer prevention trial. The data were analyzed first by a zero-truncated Poisson model and then by a zero-truncated generalized Poisson model. Use Stata's `expand` command to take the data from table 3.1, and produce one observation per woman.

   a. Compare the two models using the DIC calculated by WinBUGS.
   b. Compare the two models using the BF calculated in Stata.
   c. What prior belief would you need to have in the Poisson model to still prefer it after seeing these data?

2. The StatLib data archive includes a much analyzed dataset relating mortality rates to measures of socio-economic status and air pollution. The data can be found at http://lib.stat.cmu.edu/datasets/ under the name "pollution".

   Treat the mortality rate as the response variable and the other 15 variables as explanatory in a normal errors regression. Set your own priors for the regression coefficients. Setting priors for coefficients in a regression is sometimes easier if the explanatory variables are standardized by subtracting their mean and dividing by their standard deviation. Write a Stata program with a loop within which the program calls WinBUGS to fit each explanatory variable singly and that then extracts the DIC from the log file and finally ranks the explanatory variables in terms of model fit. Compare the ranking with that obtained from the use of p-values from Stata's `regress` command. Generalize your Stata program so that it produces forward stepwise variable selection stopping when the improvement in DIC is less than 5.

11 Further case studies

11.1 Introduction

This chapter contains four further case studies involving larger datasets and more advanced use of WinBUGS and the associated Stata routines. The case studies have been selected to illustrate specific points, so in the interest of space, some details of the analyses are only referred to briefly. All the examples are based on data used to illustrate analyses in articles published in the *Journal of Applied Statistics*, although the models used here are not identical to those in the original articles. The datasets are available at http://onlinelibrary.wiley.com/journal/10.1111/(ISSN)1467-9876/homepage/series_c_datasets.htm and relate to the articles by Bray (2002), Oman, Meir, and Haim (1999), Delmar et al. (2005), and Duchateau et al. (2003).

11.2 Case study 12: Modeling cancer incidence

Most cancer registries publish annual data on the numbers of new cases of each type of cancer in their catchment population, usually categorizing the data by age and gender. Many of these registries now have records going back over 40 or 50 years, so it is possible to study the way in which the incidence of the disease has varied by age, by period (year of observation), and by cohort (year of birth of the subjects).

Age–period–cohort (APC) models seek to create a combined description of all three time trends but are constrained by identifiability because `period = cohort + age`. Early APC models treated the coefficients of the three time scales as categorical and modeled the number of new cases, y_{ijk}, in a given number of person-years of observation, t_{ijk}, by a Poisson distribution,

$$y_{ijk} \sim \text{Poisson}(t_{ijk}\mu_{ijk})$$
$$\log(\mu_{ijk}) = \alpha_i + \beta_j + \gamma_k$$

where $i = 1, \ldots, A$ denotes the age groups, $j = 1, \ldots, P$ denotes periods, and $k = 1, \ldots, C$ denotes cohorts. This model is only identifiable if constraints are placed on the sets of coefficients, α_i, β_j, and γ_k.

One problem with the categorical APC models is that the trends can be erratic with improbable jumps in the coefficients between adjacent years. To avoid this, many researchers now use models that smooth the trends. For instance, Rutherford, Lambert, and Thompson (2010) described **apcfit**, a Stata ado-file for fitting an APC model to cancer registry data that uses splines to give smooth time trends. To use **apcfit**, you also need to install **rcsgen**, a routine that is called by **apcfit**. In this formulation, the splines are fit by maximum likelihood using software designed for generalized linear models. Earlier, Bray (2002) had described a similar Bayesian APC model that uses an autoregressive process to smooth the trends. She also provided sample data and BUGS code for fitting her model. Similar autoregressive smoothing is applied to an age–cohort model in one of the WinBUGS worked examples called ICE. The two methods of smoothing use slightly different constraints on the model parameters, so we will reparameterize Bray's Bayesian model to make it more comparable with **apcfit** and update the code from BUGS to WinBUGS syntax. For illustration, both the spline model and the adapted Bayesian model will be fit to data given by Bray (2002) on the number of new cases of cancer of the larynx in Poland.

Figure 11.1 shows the results of using apcfit on the data on cancer of the larynx in men from Poland. The data are categorized into 6 five-year periods running from 1965–1969 to 1990–1994 and 9 age groups, giving a total of 54 observations and 14 cohorts. The figure shows a rapid growth in the rate of cancer with age and a drop in the relative incidence in the more recent cohorts. The data consist of five Stata variables, cases, pyrs, a, p, and c, where the last three refer to the age, period, and cohort categories. The Stata commands that created the plot are

begin: using apcfit

```
apcfit if centre == "Poland" & gender == "male",          ///
   age(a) period(p) cohort(c) cases(cases) pop(pyrs)      ///
   agefit(afit) cohfit(cfit) perfit(pfit) nper(100000) refc(7)
twoway (rarea afit_lci afit_uci a, col(gs12) lcol(gs12))  ///
   (line afit a, lpat(solid) lcol(black)), leg(off)       ///
    xtitle(Age) ytitle("Rate per 100,000") xlabel(none) nodraw
graph save plot1.gph, replace
sort c
generate pc = p + 15
twoway (rarea pfit_lci pfit_uci pc, col(gs12) lcol(gs12)) ///
 (line pfit pc, lpat(solid) lcol(black))                  ///
 (rarea cfit_lci cfit_uci c, col(gs12) lcol(gs12))        ///
 (line cfit c, lpat(solid) lcol(black)), leg(off)         ///
 xtitle("       Cohort                          Period") ///
 ytitle(Rate Ratio) xlabel(none) nodraw
graph save plot2.gph, replace
graph combine plot1.gph plot2.gph
```

end: using apcfit

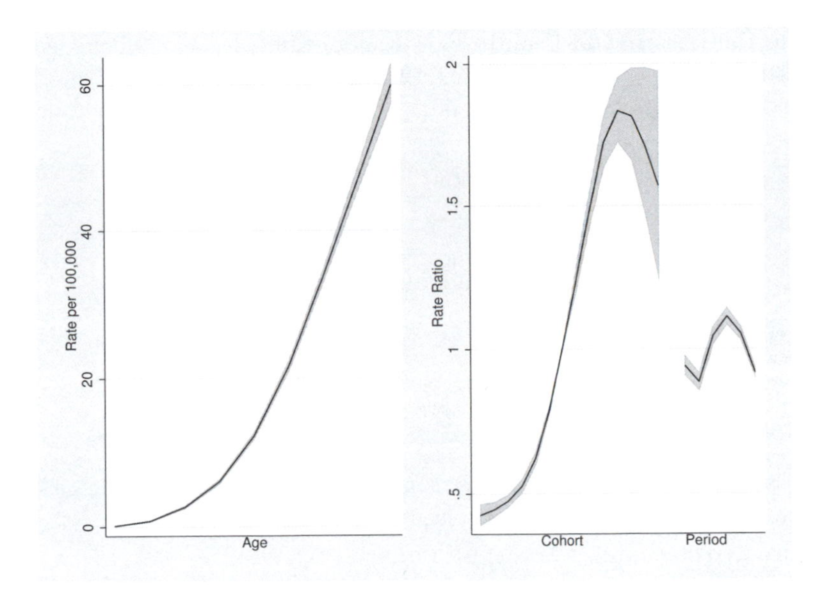

Figure 11.1. Polish males: Age, period, and cohort effects from `apcfit`

The splines used by `apcfit` place continuous curves through the data points and enable the incidence to be estimated at any age and not just in age bands. In contrast, the Bayesian approach keeps the age bands but smooths the coefficients to avoid wild fluctuations in incidence between neighboring time periods. Consider the set of age coefficients, α_1, α_2, ..., α_9. In this particular autoregressive model, the ith coefficient is modeled as depending on its immediate predecessors in such a way that

$$p(\alpha_i|\alpha_{i-1}, \alpha_{i-2}) \sim N(2\alpha_{i-1} - \alpha_{i-2}, \sigma)$$

That is, the mean for the coefficient in age band i is obtained by extending a straight line through the previous two coefficients. This process requires α_1 and α_2 to get started, so these could be given vague priors, perhaps

$$p(\alpha_i) \sim N(0, c\sigma) \quad i = 1, 2$$

where c is some arbitrary large number.

To aid convergence, we place some constraints on the sets of coefficients for age, period, and cohort. To give comparability with `apcfit`, we center the period and cohort parameters on zero and remove any linear trend in the period coefficients. Centering on zero is easy because WinBUGS has a `mean()` function, which makes it possible to write

```
for (i in 1:6) {
   gammac[i] <- gamma[i] - mean(gamma[1:6])
}
```

where `gamma` is the uncentered value obtained from the autoregressive process, and `gammac` is its centered counterpart used in the linear predictor.

Removing the linear trend is only slightly more difficult. If P coefficients, β_j, are equally spaced at points, $x_j = j - (P+1)/2$, then the average of the x's will be zero, and the slope of the regression line will be $\sum \beta_j x_j / \sum x_j^2$. Because algebraically

$$\sum_{j=1}^{P} x_j^2 = \frac{P(P+1)(P-1)}{12}$$

it is possible to remove the trend in WinBUGS by coding

```
for (p in 1:P) {
    ivec[p] <- p - (P+1)/2
    pivec[p] <- ivec[p] * beta[p]
    betac[p] <- beta[p] - ivec[p] * sum(pivec[])/(P*(P+1)*(P-1)/12)
}
```

where `beta` is the coefficient that comes from the autoregressive process, and `betac` is the coefficient with the linear trend removed.

The full WinBUGS model file takes the form

begin: APC model file

```
model
{
   for (i in 1:I) {
     cases[i] ~ dpois(mu[i])
     log(mu[i]) <- log(pyr[i]) + alpha[age[i]] + betac[period[i]]
     + gammac[cohort[i]]
   }

# age effects
   alphamean[1]    <- 0.0
   alphaprec[1]    <- taua*1.0E-6
   alphamean[2]    <- 0.0
   alphaprec[2]    <- taua*1.0E-6
   for (a in 3:A) {
      alphamean[a]    <- 2*alpha[a-1] - alpha[a-2]
      alphaprec[a]    <- taua
   }

   taua ~ dgamma(1.0E-3,1.0E-3)

   for (a in 1:A) {
      alpha[a]~ dnorm(alphamean[a],alphaprec[a])
   }

# period effects
   betamean[1]    <- 0.0
   betaprec[1]    <- taup*1.0E-6
   betamean[2]    <- 0.0
   betaprec[2]    <- taup*1.0E-6
   for (p in 3:P) {
      betamean[p]    <- 2*beta[p-1] - beta[p-2]
      betaprec[p]    <- taup
   }
```

```
      for (p in 1:P) {
         beta[p]~ dnorm(betamean[p],betaprec[p])
      }

   # constraint - mean = 0
      for (p in 1:P) {
         betam[p]<-beta[p]-mean(beta[1:P])
      }
   # constraint - linear trend = 0
      for (p in 1:P) {
         ivec[p] <- p - (P+1)/2
         pivec[p] <- ivec[p] * betam[p]
         betac[p] <- betam[p] - ivec[p] * sum(pivec[])/(P*(P+1)*(P-1)/12)
      }

      taup ~ dgamma(1.0E-3,1.0E-3)

   # cohort effects
      gammamean[1]    <- 0.0
      gammaprec[1]    <- tauc*1.0E-6
      gammamean[2]    <- 0.0
      gammaprec[2]    <- tauc*1.0E-6
      for (c in 3:C) {
         gammamean[c]    <- 2*gamma[c-1] - gamma[c-2]
         gammaprec[c]    <- tauc
      }

   # constraint: mean = 0
      for (c in 1:C) {
         gamma[c]~ dnorm(gammamean[c],gammaprec[c])
         gammac[c]<-gamma[c]-mean(gamma[1:C])
      }

      tauc~ dgamma(1.0E-3,1.0E-3)
   }
```

end: APC model file

The model can be fit in WinBUGS and a plot of the trends created by using the Stata commands:

begin: APC fitting

```
   generate pyr = pyrs/100000
   rename a age
   rename p period
   rename c cohort
   count
   local I = r(N)
   summarize age
   local A = r(max)
   summarize period
   local P = r(max)
   summarize cohort
   local C = r(max)
   * DATA FILE
   wbslist (vector cases pyr age period cohort, ///
     format(%5.0f %8.3f %3.0f %3.0f %3.0f))     ///
     (I=`I´,A=`A´,P=`P´,C=`C´) using data.txt, replace
```

```
* INITS FILE
wbslist (alpha=c(`A'{0}),beta=c(`P'{0}), ///
  gamma=c(`C'{0}),taua=100,tauc=100,taup=100) using inits.txt, replace
* RUN OPENBUGS
wbsscript using script.txt, replace                    ///
  data(data.txt) model(model_apc.txt) init(inits.txt) ///
  burnin(1000) update(10000) log(logm.txt)             ///
  set(alpha betac gammac taua tauc taup) coda(braym) openbugs
wbsrun using script.txt, openbugs
type logm.txt
wbscoda using braym, clear openbugs
```

end: APC fitting

In this analysis, the initial values were poorly chosen. To compensate for this, we had to run a relatively long burn-in of 10,000 iterations. After that, the chain suffered from high autocorrelation, and an even longer run was needed to achieve convergence. Here a run of 100,000 was thinned by 10 and took about 3 minutes to complete. Details of the convergence checking are omitted but consisted of trace and cusum plots of individual parameters and separate Mahalanobis plots of the sets of age, period, and cohort parameters with Geweke tests on the corresponding Mahalanobis distances. The Stata commands given above include the code needed to produce figure 11.2, which is similar to that obtained from `apcfit`. The Bayesian analysis uses vague priors but has the potential to incorporate further information either in the form of informative priors or via hierarchical distributions that link the trends in similar countries or in similar cancers.

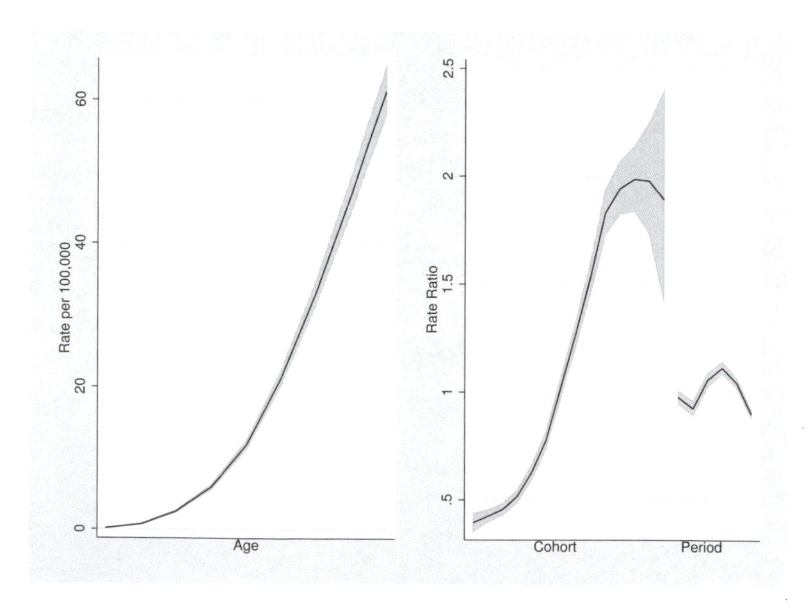

Figure 11.2. Polish men: Age, period, and cohort effects using the Bayesian model

When the data on women from Poland are analyzed in the same way, the corresponding code produces figure 11.3. Notice that the scale is different from that of the men because this cancer is much less common in women. The figure also fails to show the downturn in the more recent cohorts observed in the men. There are several possible explanations for this; for instance, the small number of cases in women might mean that the downturn is lost in the natural Poisson variation, or perhaps because the biggest risk factor for cancer of the larynx is smoking, it might be that the downturn in men is due to a change in smoking habits that has not yet taken place in women.

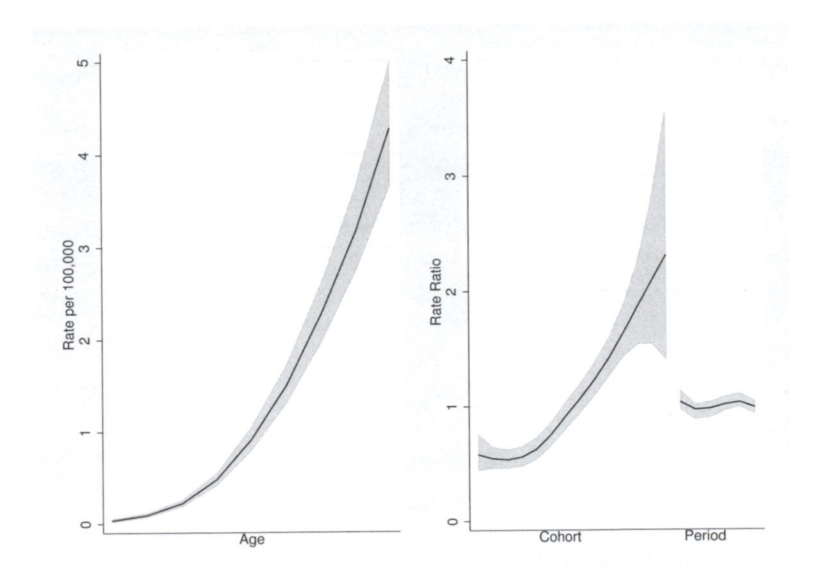

Figure 11.3. Polish women: Age, period, and cohort effects using the Bayesian model

To investigate the gender effect, we fit an extended model in which both men and women were analyzed together using the model

$$y_{ijk} \sim \text{Poisson}(t_{ijk}\mu_{ijk})$$
$$\log(\mu_{ijk}) = \alpha_i + \beta_j + \gamma_k + \theta \text{sex}_{ijk}$$

where the variable `sex` takes the value 1 for women and 0 for men. This model produces a deviance information criterion (DIC) of 915.9, while when the men were analyzed alone, the DIC was 503.4; when the women were analyzed alone, the DIC was 359.7. The total DIC when the age, cohort, and period effects are not assumed common is thus 863.1, which represents an improvement of 52.8 over the model that forced the effects to be common and shows that it would be misleading to combine the data on men and women in this way.

11.3 Case study 13: Creatinine clearance

It is important to monitor renal function regularly during chemotherapy because the drugs used can damage the kidneys. The gold standard index of kidney function is the gomerular filtration rate, but in clinical practice, it is more usual to approximate this by using the measured creatinine clearance (MCC). Creatinine is produced at a fairly steady rate by the muscles and is excreted through the kidneys. The MCC compares the amount of creatinine in the blood to that in the urine and so reflects the efficiency of the kidneys. However, accurate measurement requires the collection of a 24-hour urine sample, which is inconvenient for the patient. An alternative, but potentially less accurate index, is the estimated creatinine clearance (ECC), which replaces the 24-hour urine sample by an estimate based on the subject's weight and age.

Oman, Meir, and Haim (1999) compared MCC and ECC using data on 110 patients being treated with cisplatin chemotherapy at Hadassah Hebrew University Hospital. Measurements were taken before the start of treatment and then at intervals of about two weeks. In total, there were 253 pairs of measurements (MCC and ECC) on 60 females and 184 pairs of measurements on 50 males. Apart from gender, Oman, Meir, and Haim (1999) considered one other covariate, whether at anytime during treatment, there was evidence of a buildup of third-space body fluid, that is, fluid collecting in unusual places such as the peritoneal cavity.

The model used by Oman, Meir, and Haim (1999) was

$$
\begin{aligned}
x_{ij} &= \log_{10}(\text{MCC}) = \nu_i + \delta_{ij} \\
\nu_i &\sim N(\mu_\nu, \sigma_\nu) \\
\delta_{ij} &\sim N(0, \sigma_\delta) \\
y_{ij} &= \log_{10}(\text{ECC}) = \alpha + \beta\nu_i + \omega_i + \epsilon_{ij} \\
\omega_i &\sim N(0, \sigma_\omega) \\
\epsilon_{ij} &\sim N(0, \sigma_\epsilon) \\
\text{cov}(\delta_{ij}, \epsilon_{ij}) &= \sigma_{\delta\epsilon}
\end{aligned}
$$

where i denotes the subject, j denotes the repeat measurement, and apart from δ_{ij} and ϵ_{ij}, all terms are assumed independent. This is an example of a linear structural errors-in-variables model. Oman, Meir, and Haim (1999) fit the model using moment estimation and obtained standard errors by using the bootstrap. This approach is not without its problems because it can produce negative variance estimates, so we will consider a Bayesian analysis of the same data.

Figure 11.4 shows the relationship between individual readings of MCC and ECC on a log10 scale. It does appear that ECC overestimates the creatinine clearance, especially when the level is low.

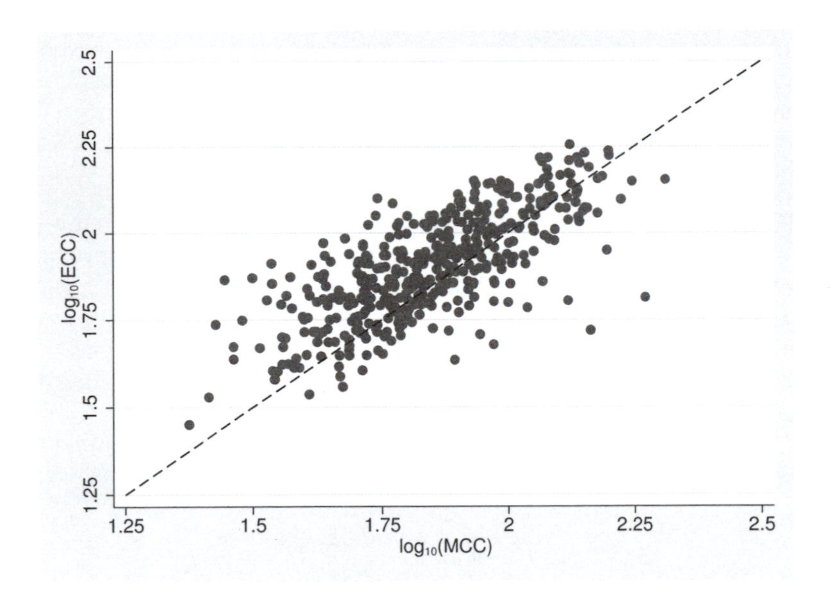

Figure 11.4. $\log_{10}$-scale plot of ECC versus MCC

Because the model is essentially a regression, the convergence of the Gibbs sampler will be improved if the covariates are centered. As can be seen from figure 11.4, both scales are centered at about 1.85, so it makes sense to work in terms of

$$\begin{aligned} x_{ij} &= \log_{10}(\text{MCC}) - 1.85 = \nu_i + \delta_{ij} \\ y_{ij} &= \log_{10}(\text{ECC}) - 1.85 = \alpha + \beta\nu_i + \omega_i + \epsilon_{ij} \\ \nu_i &\sim N(\mu_\nu, \tau_\nu) \\ \omega_i &\sim N(0, \tau_\omega) \\ (\delta_{ij}, \epsilon_{ij}) &\sim \text{MVN}_2(0, \Omega) \end{aligned}$$

where the within-subject variation of the two measurements taken at the same time point is modeled by a bivariate normal distribution (MNV_2) with precision matrix Ω, and the between-subject normal distributions have been parameterized in terms of their precisions, τ_ν and τ_ω.

On the centered scale, we would expect μ_ν and α to be close to 0, and β to be close to 1. However, before priors can be placed on the precisions, we must have some feel for the scale of measurement. Creatinine clearance is measured in units of mL/min and in healthy males ranges between about 90 and 140; women typically have values that are about 10 mL/min lower. In impaired kidney function, these values can drop to as low as 30 mL/min, but anything below that figure would usually be described as kidney failure. Without looking at the current data, we might reasonably expect the measurements to range between about $\log_{10}(30) - 1.85 = -0.37$ and $\log_{10}(140) - 1.85 = 0.30$. Thus their standard deviation should be about 0.15, and this will be made up of variation between subjects and variation within subjects. A standard deviation of 0.1 corresponds to a precision of 100, and this might be a reasonable prior guess for the mean of τ_ν and τ_ω. To allow for our uncertainty in these guesses, we will use gamma priors with the form $G(2, 50)$, or in the WinBUGS parameterization, we will use gamma priors with the form $G(2, 0.02)$.

The Wishart prior on the joint within-subject variation will be given a low number of degrees of freedom, k, to reflect extreme uncertainty. A crude approximation says that if you want the ratio of the standard deviation of the precision to the mean of the precision to be r, then you should use $k = 2/r^2$. So using $k = 2$ implies that the mean and standard deviation of the precision are expected to be of similar size. WinBUGS parameterizes the Wishart in terms of a matrix R such that the expected precision matrix is kR^{-1}, or put another way, the variance matrix should be close to R/k. A precision of 100 implies a variance of 0.01, and if we assume a correlation of 0.5 between the variation on the two scales, then

$$R = \begin{pmatrix} 0.02 & 0.01 \\ 0.01 & 0.02 \end{pmatrix}$$

The priors that will be used are, in WinBUGS parameterization,

$$\begin{aligned} a &\sim N(0, 1) \\ b &\sim N(1, 1) \\ m_v &\sim N(0, 1) \\ t_v &\sim G(2, 0.02) \\ t_w &\sim G(2, 0.02) \\ \Omega &\sim \text{Wishart}(R, 2) \end{aligned}$$

and the model file is

begin: MCC–ECC model file

```
model {
   for (i in 1:437) {
      x[i,1:2] ~ dmnorm(mu[i,1:2],Omega[1:2,1:2])
      mu[i,1] <- v[id[i]]
      mu[i,2] <- a + b*mu[i,1] + w[id[i]]
   }
   for (i in 1:110) {
      v[i] ~ dnorm(mv,tauv)
      w[i] ~ dnorm(0,tauw)
   }
   Omega[1:2,1:2] ~ dwish(R[1:2,1:2],2)
   tauv ~ dgamma(2,0.02)
   tauw ~ dgamma(2,0.02)
   a ~ dnorm(0,1)
   b ~ dnorm(1,1)
   mv ~ dnorm(0,1)
```

```
# PREDICTIONS FOR MODEL CHECKING

   for (i in 1:110) {
      vstar[i] ~ dnorm(mv,tauv)
      wstar[i] ~ dnorm(0,tauw)
   }
   for (i in 1:437) {
      xstar[i,1:2] ~ dmnorm(mustar[i,1:2],Omega[1:2,1:2])
      bstar[i,1:2] ~ dmnorm(mstar[i,1:2],Omega[1:2,1:2])
      mustar[i,1] <- v[id[i]]
      mustar[i,2] <- a + b*v[id[i]] + w[id[i]]
      mstar[i,1]  <- vstar[id[i]]
      mstar[i,2]  <- a + b*vstar[id[i]] + wstar[id[i]]
   }
}
```

end: MCC–ECC model file

For model checking, it is not necessary to store every cycle of the chain. If every cycle were stored, it would slow down the analysis to no benefit. There are 437×2 observations in the dataset; therefore, there are 437×2 potential predictions that could be made. If we were to store a chain of length 10,000, those predictions would create a file with 8.74 million values. Both WinBUGS and Stata could cope with such an analysis, but it would be slow, especially when it comes to reading the Coda files into Stata.

A sensible strategy is to run a long chain to access convergence and estimate the parameters but then to thin the chain to a length of about 1,000 to assess model fit. Such a thinned chain will have under a million predictions so will be quicker to process but will still be sufficiently accurate for model checking. In fact, this model converges quickly, and so model checking was based on a burn-in of 1,000 followed by a run of 10,000 that was thinned by 10.

The results are

```
. mcmcstats a b mv sdv sdw sdd sde rho
```

Parameter	n	mean	sd	sem	median	95% CrI
a	10000	0.060	0.008	0.0003	0.060	(0.045, 0.077)
b	10000	0.824	0.068	0.0030	0.822	(0.693, 0.959)
mv	10000	0.002	0.014	0.0002	0.002	(-0.025, 0.029)
sdv	10000	0.128	0.011	0.0001	0.127	(0.108, 0.150)
sdw	10000	0.073	0.006	0.0001	0.072	(0.061, 0.086)
sdd	10000	0.107	0.004	0.0001	0.107	(0.099, 0.116)
sde	10000	0.070	0.003	0.0000	0.070	(0.065, 0.076)
rho	10000	0.544	0.039	0.0005	0.545	(0.465, 0.617)

This model has two types of random effects and thus has two sorts of residuals. At the levels of d_{ij} and e_{ij}, there are the residuals of the individual measurements about the estimated mean for that subject; at the levels of ν_i and ω_i, there are the residuals between the mean for each subject and the estimated mean for all subjects. Each set of residuals tells us something different about the model. In the WinBUGS code, the individual observations are compared with **xstar**, which only resamples d_{ij} and e_{ij}, and the subject means are compared with the corresponding average **bstar**, which also resamples ν_i and ω_i.

The code below implements these model checks but starts by omitting one subject with extremely unusual observations to mirror the analysis in Oman, Meir, and Haim (1999).

begin: residual plots

```
* remove outlier and centre

use oman.dta, clear
replace id = id + 61*(sex==2)
drop if id == 61
replace id = id - 1 if id > 61
generate x_1 = log10(MCC) - 1.85
generate x_2 = log10(ECC) - 1.85
sort id
save omanraw.dta, replace

* Write data, initial values & script then run

matrix R = 0.02,0.01\ 0.01,0.02
wbslist (vector id,  format(%4.0f)) (matrix R) ///
    (structure x_1 x_2, name(x) format(%8.3f))  using data.txt, replace
wbslist (Omega=structure(.Data=c(130,-65,-65,130),.Dim=c(2,2)), ///
   mv=0,a=0,b=1,tauw=100,tauv=100) using inits.txt, replace
wbsscript using script.txt, replace ///
  data(data.txt) model(model.txt) init(inits.txt) ///
  burnin(1000) update(10000) thin(10) dic logfile(log.txt) ///
   set(a b mv tauw tauv Omega xstar bstar) coda(oman) openbugs
wbsrun using script.txt, openbugs

* Read the predictions for individual readings

wbscoda using oman, clear openbugs keep(xstar*)

* Recode xstar_17_1 as xstar_1_17

rename xstar_#_# xs_#[2]_#[1]
save mcmc.dta, replace

* Residual plots

mcmccheck, d(x_1) pred(xs_1_) dfile(omanraw.dta) pfile(mcmc.dta) ///
  plot(summary) cgopt(title(MCC))
mcmccheck, d(x_2) pred(xs_2_) dfile(omanraw.dta) pfile(mcmc.dta) ///
  plot(summary) cgopt(title(ECC))

* read predictions for subject means & recode

wbscoda using oman, clear openbugs keep(bstar*)
rename bstar_#_# bs_#[2]_#[1]
save mcmcb.dta, replace
```

```
* Observed subject means

use omanraw.dta, clear
sort id
by id : egen mx_1 = mean(x_1)
by id : egen mx_2 = mean(x_2)
keep if id != id[_n-1]
save omanmean.dta, replace

* Corresponding mean predictions

use omanraw.dta, clear
forvalues i=1/437 {
  local id`i´ = id[`i´]
}
use mcmcb.dta, clear
forvalues j=1/110 {
  local vlist1 ""
  local vlist2 ""
  forvalues i=1/437 {
     if `id`i´´ == `j´ {
        local vlist1 "`vlist1´ bs_1_`i´"
        local vlist2 "`vlist2´ bs_2_`i´"
     }
  }
  egen mbs_1_`j´ = rmean(`vlist1´)
  egen mbs_2_`j´ = rmean(`vlist2´)
}
keep mbs*
save mcmcb.dta, replace

* Residual plots

mcmccheck, d(mx_1) pred(mbs_1_) dfile(omanmean.dta) pfile(mcmcb.dta) ///
  plot(summary) cgopt(title(MCC))
mcmccheck, d(mx_2) pred(mbs_2_) dfile(omanmean.dta) pfile(mcmcb.dta) ///
  plot(summary) cgopt(title(ECC))
```

end: residual plots

The summary residual plots for $\log_{10}$(ECC) are shown as figure 11.5; those for MCC are similar. The upper plots show the residuals based on the 437 individual readings and highlight the way that these values vary about the subjects' mean predicted levels. They are unremarkable except that the variance does appear to be greater at lower levels of creatinine clearance. This is connected to the pattern that we observed in figure 11.4. The lower four plots show the variation in the subjects' mean levels. These are more or less as one would expect for normally distributed between-subject variation. The model has no covariates, so the predicted overall mean of $\log_{10}$(ECC) is $\alpha + \beta\mu_\nu$ for every subject, and the variation seen in the mean predictions is due to sampling error in the Markov chain Monte Carlo (MCMC) algorithm. Nonetheless, the residuals are of the size that one would expect, and overall, the model seems to fit quite well.

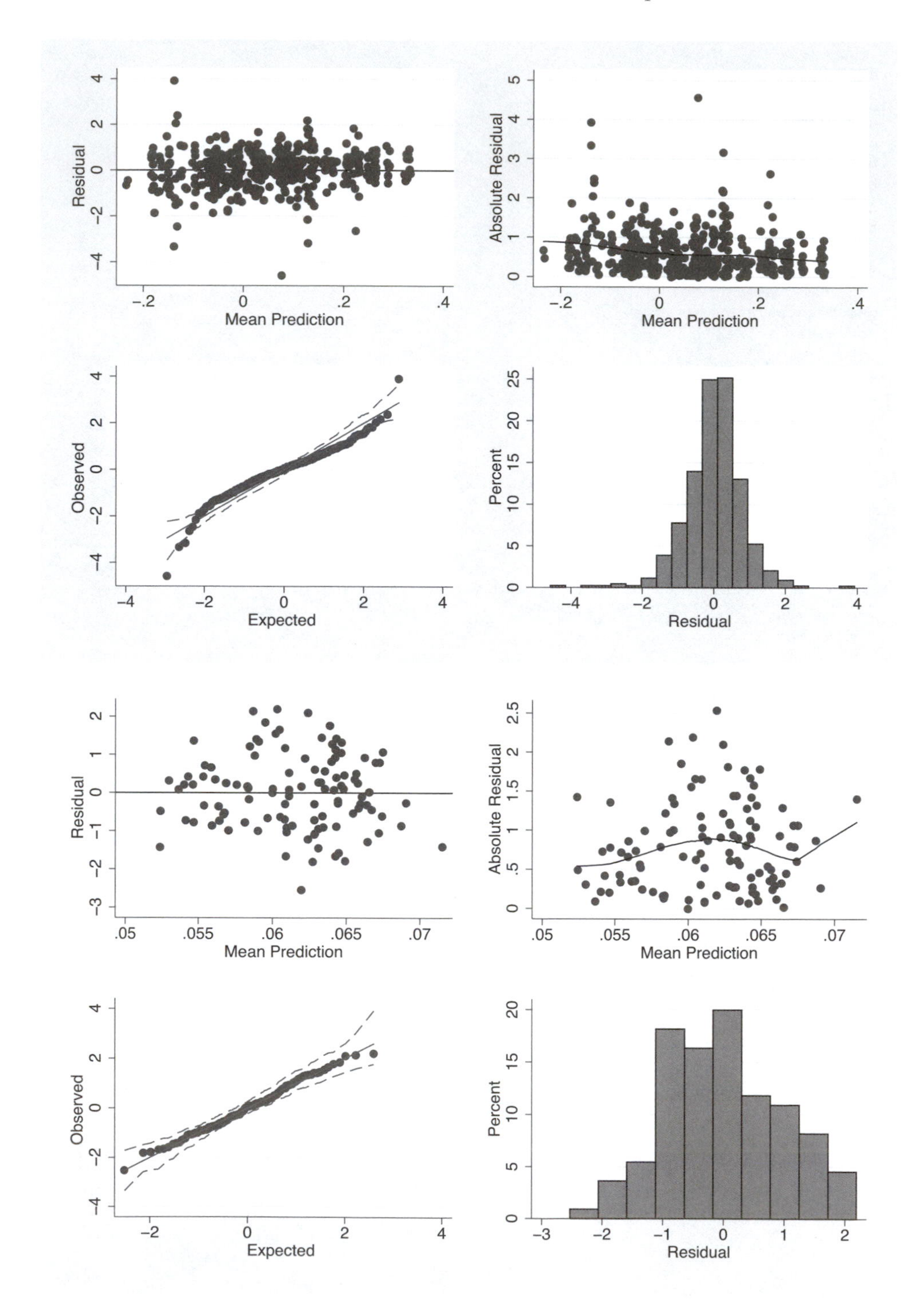

Figure 11.5. Summary residual plots for $\log_{10}$(ECC); upper plot shows individual residuals, and lower plot shows residuals based on subject means

Prior knowledge tells us that the levels of creatine clearance for men and women might be different, although this does not necessarily imply that the relationship between MCC and ECC varies with gender. As Oman, Meir, and Haim (1999) analyzed separately by gender and fluid retention, it makes sense to investigate this by creating four groups and fitting separate regressions within each group by allowing α, β, and μ_ν to vary by group. We leave the variance structure to be common across groups and use the same priors. As a third analysis, a model was fit in which μ_ν varies by group but α and β are common. The DIC for the model presented above that did not distinguish between groups was -1771.0; this improved by 5.0 in the third model in which the parameters μ_ν were allowed to vary by group, which suggests that creatinine clearance does indeed vary with gender and third body fluids. However, as we have seen, the subject mean levels vary little, and although these effects may be genuine, they are small. When in the second model, the relationship between MCC and ECC was also allowed to vary by group through the introduction of group-specific parameters α and β, the DIC got slightly worse. There does not seem to be any reason to believe that the calibration of $\log_{10}(\text{MCC})$ on $\log_{10}(\text{ECC})$ varies with gender or with fluid retention, even though the levels vary slightly by these factors.

11.4 Case study 14: Microarray experiment

Genes are sections of DNA, most of which include the code for creating their own specific protein. Now although DNA is present in cells throughout the body, the genes within the DNA are not always switched on; that is, different tissues will create particular proteins at particular times, and microarrays are used to measure this pattern of gene expression.

Microarrays consist of a plate covered with probes made from sequences of DNA corresponding to specific genes. When a gene is switched on to produce its protein, it must first create an RNA, which acts as a template on which that protein is constructed. In a microarray experiment, all the different RNAs taken from a tissue sample are processed with enzymes to produce complementary DNA (cDNA); this has the property that it will hybridize with the corresponding probes on the microarray. The more RNA of each type that the gene has produced, the more cDNA there will be to stick to the probe. In this way, it is possible to quantify the expression of each gene.

In a two-color microarray experiment, tissue samples are taken from two sources, and the cDNA is labeled with different colored dyes. The mixed sample is then hybridized to a microarray, and the intensities of the two colors are measured at each probe. The comparative expression is usually quantified as $\log_2$ of the ratio of the intensities of the two colors for each gene.

Delmar et al. (2005) analyzed a two-color microarray experiment in which they compared the genes expressed by the spleens of control mice with those expressed in spleens taken from mice three hours after they had been exposed to 1 Gy of whole-body irradiation. The microarray that they used had probes for 4,360 genes, and the experiment was performed as 3 dye swaps; that is, for each of the three pairs of control and irradiated tissue samples, the two colored dyes were swapped between one replicate and the other. Thus in total, there were six replications. With these six replicates, the object of the analysis was to detect any genes that were more highly, or less highly, expressed as a result of the irradiation.

Microarray experiments are typified by the unfortunate combination of a very large number of outcome measures and a very small number of replicates. One simple approach to their analysis is to treat each gene separately and to perform some statistical test, such as a one-sample t test, to see whether the log of the ratio of intensities is different from zero. In this way, the genes can be ranked by their p-values. Unfortunately, this approach is hampered by the instability of the variance estimates that results from the small number of replicates; thus the test tends to have low power.

Many methods have been suggested for improving the performance of the t test by stabilizing the estimates of the between-replicate variance, typically by combining the variance estimate for each specific gene with some average variance calculated from the entire set of genes. The approach adopted by Delmar et al. (2005) was to suppose that the genes can be divided into a small number of sets such that within each set, the variance between replicates is constant across the genes. The variance estimate from each gene is then stabilized by using the estimate derived from all members of its set. The statistical problems involved in implementing this approach include choosing how many gene sets to use and deciding how to allocate each gene to the best set. Delmar et al.'s (2005) approach is an example of a mixture model on the variance and in their article, it was fit by maximum likelihood using an expectation maximization algorithm. We will adapt their model and fit it to the mouse spleen experiment using a Bayesian approach, but first, we will consider a few simpler alternative models.

In a non-Bayesian analysis, each gene's one-sample t statistic can be calculated directly using the mean and standard error of the six replicates. With the data in Stata variables `logratio1` to `logratio6`, the t statistics can be found using the commands

```
egen m = rmean(logratio1-logratio6)
egen s = rsd(logratio1-logratio6)
generate se = s/sqrt(6)
generate t = m/se
```

Figure 11.6 shows the plot of the calculated t statistics against the expected values under a t distribution with 5 degrees of freedom. It is clear that the alignment of the t statistics is too steep, which suggests that the distribution of t statistics has wider tails than expected, and this might cause p-values from a t test to exaggerate the significance. Despite this, there is a suggestion in the extreme tails that some genes have changed their expression as a result of the irradiation.

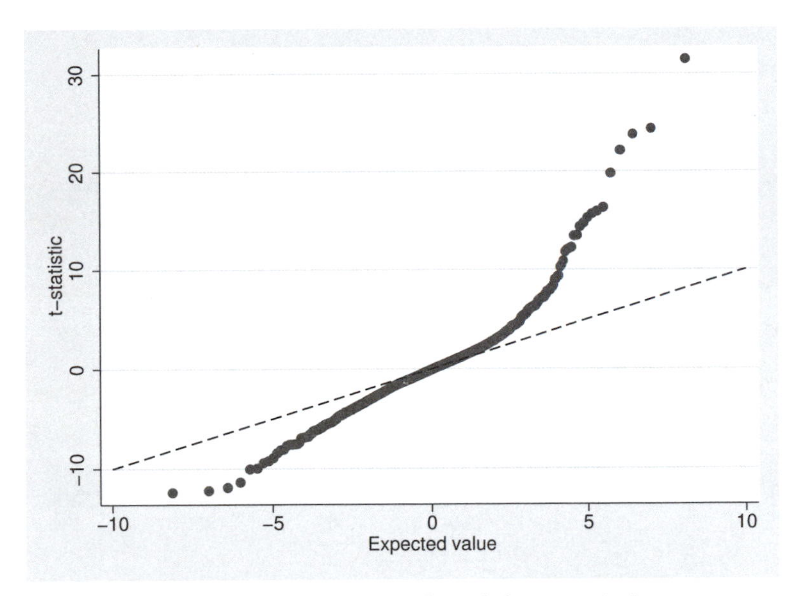

Figure 11.6. Quantile plot of the t statistics

Because one of the common problems with microarray experiments is extremes of variability, it is important to look at the distribution of the standard deviation, s, or because we are to undertake a Bayesian analysis, the precision, $1/s^2$. A histogram of the 4,360 precisions and a superimposed gamma distribution, $G(1.5, 40)$, is shown in figure 11.7. The parameterization used here is that adopted by Stata, so the mean of the distribution is $1.5 \times 40 = 60$.

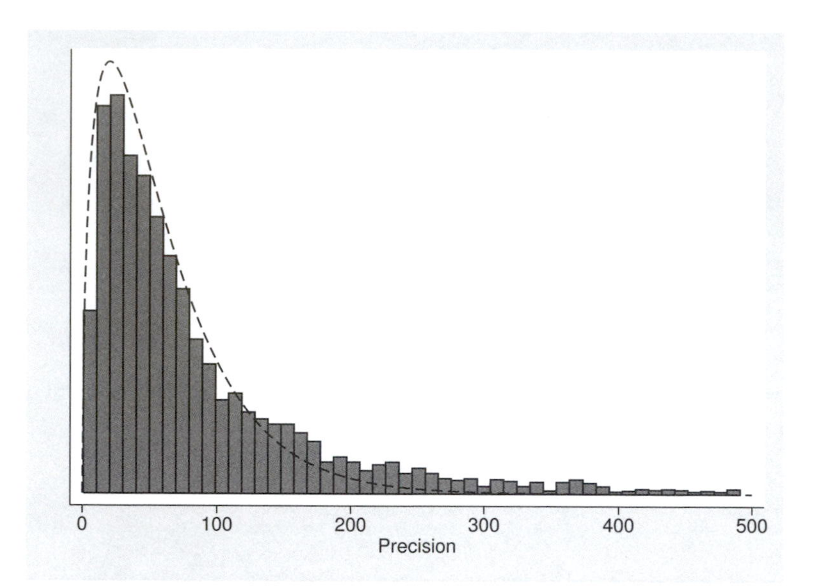

Figure 11.7. Histogram of calculated precisions for each of the 4,360 genes with a superimposed gamma, $G(1.5, 40)$, distribution

If we adopted $G(1.5, 40)$ as the prior for the precisions, then we would have a distribution that is in close agreement with the bulk of the actual observations but that would tend to draw the precision estimates for the more extreme genes back toward the average for all genes. This approach of choosing a prior based on the data has its advocates; for instance, it was suggested by Baldi and Long (2001) in an influential article on microarray analysis, but it is questionable on the grounds of using the same data twice, once to create the prior and once in the likelihood. A less controversial alternative is to create a hierarchical distribution for the precisions with the form $G(a, b)$ and then to put vague priors on a and b. With such a large set of genes, the hierarchical model inevitably produces estimates close to $a = 1.5$ and $b = 40$. In this case, the estimates of the precisions of each gene are much the same as those found under a $G(1.5, 40)$ prior. This argument justifies the adoption of the data-based $G(1.5, 40)$ prior as a sensible approximation to a hierarchical model.

A similar argument leads us to a normal distribution with mean 0 and standard deviation 0.08 for the means of the genes as shown in figure 11.8. If we adopted this distribution as the prior for the means, either directly or by assuming a $N(0, \sigma)$ hierarchical structure and estimating sigma from the full set of genes, then the effect would be that the Bayesian estimates of t would lie close to their expected values in a quantile plot similar to figure 11.6. However, because the prior is so well defined and the number of replicates for any one gene is only six, the prior would have a powerful influence on the few extreme genes, and their estimated effects would be pulled strongly toward zero, leaving us with little information about which genes were over or under expressed.

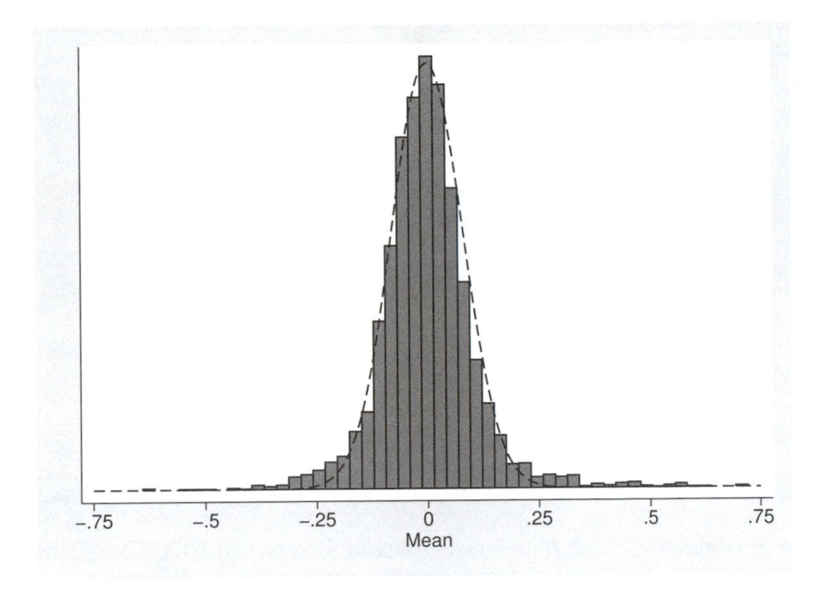

Figure 11.8. Histogram of calculated mean for each of the 4,360 genes together with a normal, $N(0, 0.08)$, distribution

The difference between the data-based prior on the precisions and the data-based prior on the means is important. In the case of the precisions, we believe that extremes are the result of chance and have no biological basis, so it makes sense to shrink extreme precisions back toward those of the rest of the sample. However, the experiment was performed because it was believed that irradiation would change the expression for a few genes, so outlying means are biologically meaningful, and we do not want to shrink those back toward zero. In the case of the means, we need to think much more carefully about the prior so that our choice truly reflects what we believe is happening; a poorly chosen prior could overwhelm the data and produce a misleading analysis.

One way to specify a more meaningful prior for the mean is to imagine a breakdown of the process by which underlying mean log ratios would be generated in a simulation. For a large proportion of the genes, we expect the irradiation to have no effect, so their means would be zero. Let p be the proportion of genes that have their expression influenced by the irradiation; we might anticipate that p will be under 0.05, although it would be good to take advice on this from experts. When there is an effect, it could be either to increase the expression or to decrease it. In the absence of more specific knowledge, we might assume that changes in either direction are equally likely. Finally, we need to quantify our expectation about the order of magnitude of the change in expression. Microarray experiments typically look for effects of the order of a 1.5- or 2-fold difference in the expression, so we might treat the expected size of any true effects as being $N(1.5, 0.5)$, the relatively large standard deviation reflecting our uncertainty.

Combining the staged prior with an assumed gamma distribution for the precision leads us to the WinBUGS model file.

begin: model file

```
model {
  for (i in 1:4360) {
     d[i] ~ dbern(p)
     s[i] ~ dbern(0.5)
     m[i] ~ dnorm(1.5,4)
     mu[i] <- d[i]*(1-2*s[i])*m[i]
     prec[i] ~ dgamma(1.5,0.025)
     for (j in 1:6) {
        y[i,j] ~ dnorm(mu[i],prec[i])
     }
     t[i] <- mu[i]*sqrt(6*prec[i])
  }
  p ~ dbeta(1,50)
}
```

end: model file

Here the parameterizations have been changed to suit WinBUGS, and the parameter p is given a beta prior with a mean of about 2%. To prepare the data, set random initial values, and fit the model, we use the following commands:

begin: model fitting

```
wbslist (structure logratio1-logratio6, name(y) format(%6.3f))  ///
    using data.txt, replace
wbslist (prec=c(4360{100}),p=0.02,s=c(4360{runiform()<0.5}),m=c(4360{1.5}), ///
   d=c(4360{runiform()<0.02})) using inits.txt, replace
wbsscript using script.txt, replace                            ///
  data(data.txt) model(modelm.txt) init(inits.txt)             ///
  burnin(1000) update(10000) thin(10) dic logfile(logm.txt) ///
  set(t p) coda(delmarm) openbugs
wbsrun using script.txt, openbugs
wbscoda using delmarm, clear openbugs
```

end: model fitting

Using a burn-in of 1,000 and a run length of 10,000 with thinning by 10 gives a saved set of 1,000 simulations. For most of the genes, the means and, hence, the t statistics were zero for most simulations. Only 20 genes were nonzero for more than 50% of the chain, and only 5 genes never had a mean of 0; these included *p21* and *BAX*, two of the genes highlighted by Delmar et al. (2005) as being known to be affected by irradiation.

A mixture model for the variance as advocated by Delmar et al. (2005) offers an interesting alternative way of stabilizing the variance, but as we have already seen, the simpler approach of using the approximate distribution of the precisions across all genes works well. The way in which the variance is stabilized is probably less important than the way in which the mean is modeled. Delmar et al. (2005) used an expectation maximization algorithm for maximum likelihood estimation, so in their analysis, there was no prior on the mean; thus the means were not shrunk toward zero. To imitate this in a Bayesian analysis, we would have to assume, rather unrealistically, that prior to the experiment, we know nothing about the expected log ratios and hence are happy to place a vague prior on them.

To create a mixture of, say, three variances, we could associate each gene with a number $k = 1, 2, 3$ and then link each value of k to a different variance σ_k^2. These variances could be estimated from the set of genes with the same k. Such models are referred to as mixtures and are notoriously difficult to fit by either likelihood or Bayesian methods. To see why, suppose that the three variances were 0.3, 0.4, and 1.0. One solution would be $\sigma_1^2 = 0.3$, $\sigma_2^2 = 0.4$, and $\sigma_3^2 = 1.0$: an equally valid solution would be $\sigma_1^2 = 0.4$, $\sigma_2^2 = 1.0$, and $\sigma_3^2 = 0.3$ but with the genes reassigned to corresponding groups; there are six such solutions in total. The likelihood under this model consists of six similar-sized discrete hills corresponding to the six solutions. When the prior is relatively flat, the posterior will have a similar shape. A well-mixing MCMC algorithm jumps freely between these different solutions, and if we could create such a freely mixing chain, the results would be difficult to interpret because the chains for σ_1^2, σ_2^2, and σ_3^2 would all have the same mean.

In practice, the MCMC chain for a mixture model rarely mixes freely; instead, it tends to find one solution, or hill within the posterior, and to stay with it. Thus Bayesian algorithms usually create a poorly mixing chain that gives us a single solution, although which solution we get will probably depend on the initial values. Of course, knowing that we will probably home in on a unique solution does not guarantee that we actually will, and it is possible that the chain will switch solutions at some point, especially if two of the solutions lie close together. It would clearly be much better if the problem were parameterized in such a way that multiple solutions were avoided, for example, by insisting through the prior that $\sigma_1^2 > \sigma_2^2 > \sigma_3^2$. One way to achieve this is to use WinBUGS's censoring operator to limit the range of the prior. An implementation of this approach for a three-component mixture is

begin: model file

```
model {
   for (i in 1:4360) {
      set[i] ~ dcat(p[1:3])
      mu[i] ~ dnorm(0,0.001)
      for (j in 1:6) {
         y[i,j] ~ dnorm(mu[i],prec[set[i]])
      }
      t[i] <- mu[i]*sqrt(6*prec[set[i]])
   }
   prec[1] ~ dgamma(1.5,0.025) C(,prec[2])
   prec[2] ~ dgamma(1.5,0.025) C(prec[1],prec[3])
   prec[3] ~ dgamma(1.5,0.025) C(prec[2],)
   p[1:3] ~ ddirich(alpha[1:3])

# log likelihood

   for (i in 1:4360) {
      for (j in 1:6) {
         LL[i,j] <- 0.5*log(prec[set[i]])
  - 0.5*prec[set[i]]*(y[i,j]-mu[i])
                 *(y[i,j]-mu[i])
      }
   }
   logL <- sum(LL[,])
```

```
# results for gene p21

   keep[1] <- mu[2090]
   keep[2] <- prec[set[2090]]
   keep[3] <- t[2090]
   keep[4] <- set[2090]
}
```

end: model file

Unfortunately, WinBUGS cannot calculate the DIC for this model, so a section of code is included that calculates the log likelihood using the formula for the density of the normal distribution. The final section of the code stores results for a gene of special interest that happens to be number 2,090 in the list of 4,360 genes. This is the *CDKN1A* gene located on chromosome 6, which codes for the *p21* protein involved in stopping cell growth when a cell has been damaged, for instance, by irradiation.

Fitting models with between 2 and 5 components in the mixture gave mean posterior logL's that increased from 36,371 to 37,046 to 37,159 to 37,160. However, the standard errors of these means were all about 2.4, so differences below about 6 or 7 could be due to sampling error in the MCMC chains. With each increasing component in the mixture, we add a further two parameters, one extra proportion, and one extra precision; thus a Bayesian version of the Akaike information criterion penalizes the difference between successive logL's by 4, and it seems clear that there is little gain in having more than 4 variances in the mixture.

The posteriors of the standard deviation for the *p21* gene under different numbers of components are shown in figure 11.9. The figure highlights the fact that the posterior mean is often a poor summary under a mixture model.

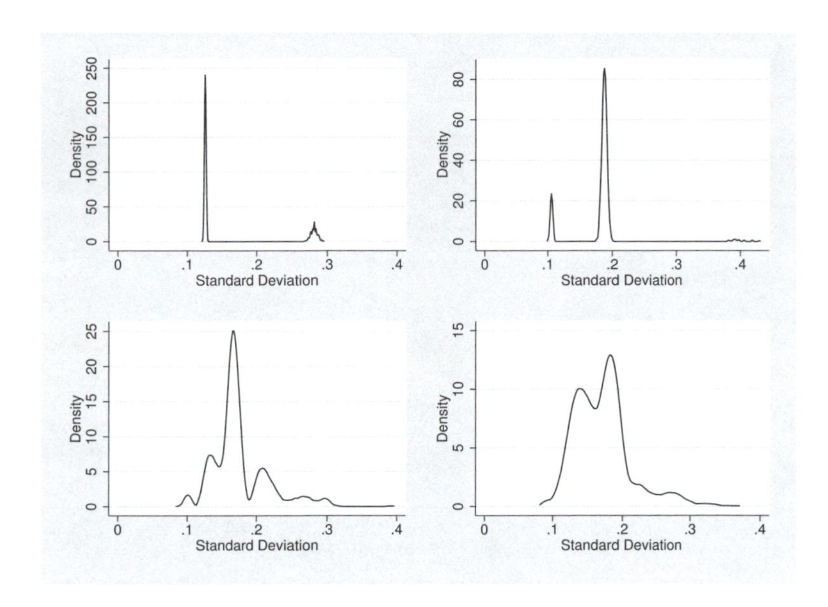

Figure 11.9. Posterior distributions of the standard deviation of the gene coding for the *p21* protein under models with 2, 3, 4, and 5 components

Finally, as a comparison of the four-component mixture model and the model that assumed a $G(1.5, 40)$ prior on the precision and a small proportion of nonzero effects, the corresponding t statistics were plotted against one another in figure 11.10. The t statistics of the most extreme genes are similar, but the nonmixture model is easier to fit and has the advantage of making it clearer when a gene is unaffected by irradiation.

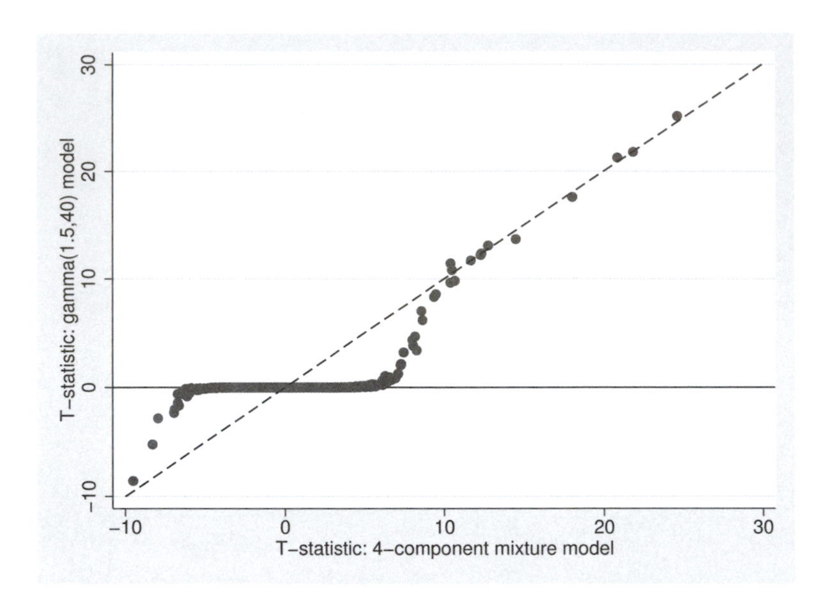

Figure 11.10. T statistics for each of the 4,360 genes under the gamma variance model that assumes a limited number of nonzero effects and the 4-component variance mixture model with a vague prior on the mean

11.5 Case study 15: Recurrent asthma attacks

Recurrent events occur in trials of treatments for diseases such as asthma when a patient may experience several separate attacks during the time of the study. It is possible to use a standard survival model for the time to the first attack, but after that, there are two possible strategies corresponding to analyses on different time scales. For the next at-risk period, we could either restart the clock and measure time from the end of the first attack or continue to measure time from the start of the trial. Which time scale is more appropriate depends on the nature of the attacks. If the risk of an attack is higher in the period following a previous attack, then it might make sense to restart the clock. On the other hand, if this is not the case and risk changes with age, then it makes more sense to measure time from the start of the study. Duchateau et al. (2003) investigated models for recurrent events using different time scales and applied them to the results of an asthma trial.

The data used by Duchateau et al. (2003) consist of information on 1,037 at-risk periods on 232 children recruited into an asthma prevention trial. The authors do not identify the trial, but the data appear to represent a subset of children from the Early Treatment of the Atopic Child (ETAC) study in which children with atopic dermatitis were randomized to receive either cetirizine or a placebo (Warner 2001). For each risk period, the dataset lists the following: `id`, a child identifier; `trt`, a treatment indicator;

start and stop, the start and end times measured in days from the start of the trial; and event, a censoring indicator denoting whether that risk period ended in an asthma attack. Figure 11.11 shows the patterns of events for the first 20 children and was created by the Stata commands

```
bysort id: generate end = start[_n+1]
twoway (rcap start stop id if id <= 20, horiz)                   ///
   (rcap stop end id if id <= 20 & event == 1 & id[_n+1]==id, ///
     horiz lwidth(thick) lcol(black))                         ///
   (rcap stop stop id if id <= 20 & event == 1 & end==.,      ///
     horiz lwidth(thick) lcol(black)),                        ///
   leg(off) xtitle(Calendar time (days)) ytitle(Child)        ///
   xlabel(0(100)700) xscale(range(0 700))
```

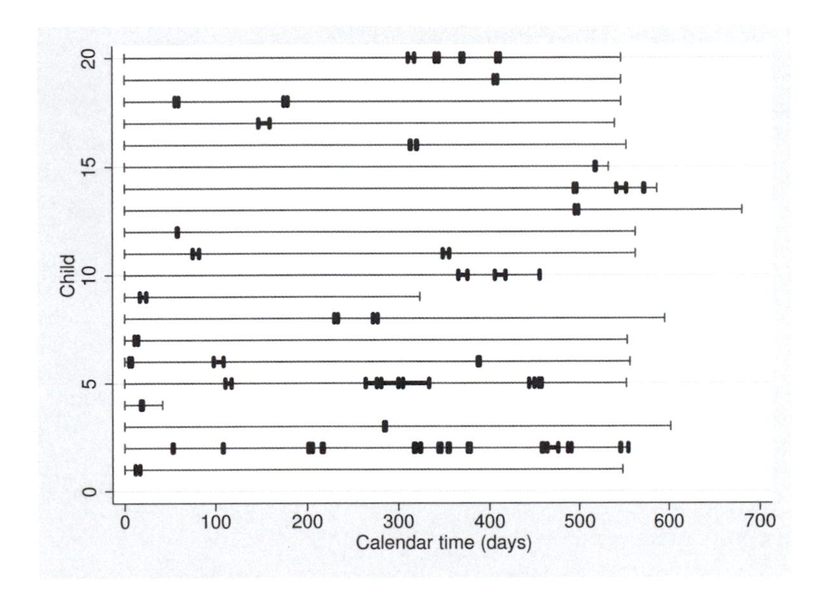

Figure 11.11. Illustrations of the pattern of recurrent events from the asthma study

In the above code, the end times of the periods when the child was not at risk are calculated from the start times of the following periods, and then the plot is created in three parts: lines are drawn for the at-risk periods; thicker lines are added for the periods when the child is not at risk because he or she are having an asthma attack; and a thick marker is added to show any events that occurred at the end of a child's last at-risk period.

Children are not at risk of a new attack until the current attack is over, but because these periods are very short and because few children drop out before 500 days, a good indication of the way that the hazard changed over calendar time can be obtained by counting the number of events in successive periods of 50 days.

```
forvalues i=1/12 {
   count if event == 1 & stop > (`i´-1)*50 & stop < `i´*50
}
```

The results are 78, 69, 68, 70, 62, 57, 80, 81, 86, and 75; it certainly appears that the hazard does not change much with calendar time. However, if the clock is restarted after each attack by calculating the gap time from the last attack, then the proportion of events in successive periods of 50 days can be found from the commands

```
generate gap = stop - start
forvalues i=1/10 {
  quietly count if gap > (`i´-1)*50
  local a = r(N)
  quietly count if event == 1 & gap > (`i´-1)*50 & gap < `i´*50
  display %4.0f r(N) %5.2f r(N)/`a´
}
```

This produces the values 0.45, 0.21, 0.19, 0.13, 0.11, 0.16, 0.17, 0.18, 0.20, and 0.29 and suggests that the risk, and thus the hazard of an asthma attack, is higher at the start of the at-risk periods and perhaps that it increases again after about a year. However, the eventual increase is based on relatively few observations and may not be genuine.

Alternatively, we could use Stata's survival analysis command `sts graph` to plot the estimated hazards. To do this, use the commands

```
stset stop, id(id) failure(event) exit(time .) time0(start)
sts graph, hazard yscale(range(0 0.01)) ///
   ylabel(0(0.002)0.01) title(Calendar time)
graph save plot1.gph, replace
stset stop, failure(event) origin(start)
sts graph, hazard yscale(range(0 0.01)) ylabel(0(0.002)0.01) title(Gap time)
graph save plot2.gph, replace
graph combine plot1.gph plot2.gph
```

to produce figure 11.12, which shows that the hazard over calendar time is quite flat, while the hazard over gap time shows a drop over the first 200 days and subsequent rise.

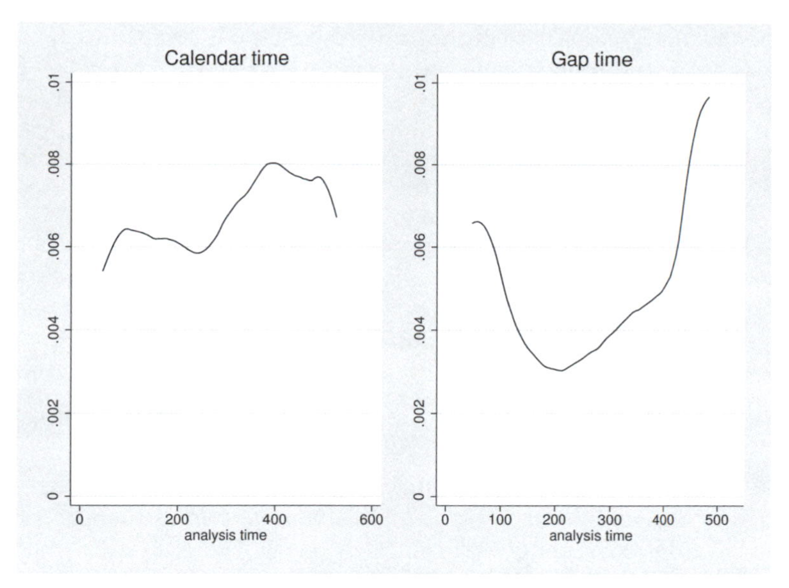

Figure 11.12. Hazard estimates over calendar and gap time

As is standard for a survival study, Duchateau et al. (2003) model the hazard, $h_i(t)$, for the ith child at time t by the following:

$$h_i(t) = h_0(t) U_i \exp(bx_i)$$

That is, for a child at risk, the probability of an asthma attack is made up of a baseline hazard, $h_0(t)$, common to all children, an exponential term that incorporates the effect of that child's covariates, and a frailty term, U_i, that describes the way in which that child is more or less at risk than the average for his or her covariates.

A popular model in survival analysis is the Weibull, which has a shape parameter, υ, and a scale parameter, λ. Under the Weibull, the hazard takes the form

$$h_i(t) = \upsilon \lambda t^{\upsilon - 1} \qquad \upsilon, \lambda > 0$$

So depending on the values of the parameters, the hazard either increases or decreases steadily over time according to some power curve. The Weibull does not allow for the u-shaped pattern seen for gap time in figure 11.12 so would be inappropriate if the rise in hazard after a year were believed to be genuine. However, if we accept the Weibull and allow the scale parameter to depend on the covariates and add a normal random effect, u_i, then a natural structure is

$$\log(\lambda) = \alpha + \alpha x_i + u_i$$

which makes the hazard

$$\begin{aligned} h_i(t) &= \upsilon \exp(\alpha + \beta x_i + u_i) t^{\upsilon - 1} \\ &= \upsilon \exp(\alpha) t^{\upsilon - 1} \exp(u_i) \exp(\beta x_i) \end{aligned}$$

This corresponds to a Weibull baseline hazard with shape υ and scale $\exp(\alpha)$, a frailty described by $\exp(u_i)$, and proportional effects for the covariates. This model differs from that used by Duchateau et al. (2003) only in that they place a gamma distribution on the frailty, U_i, while here we have a log-normal distribution. The difference is slight, but the log-normal representation is a little easier to fit.

If the random effects, u_i, are given a zero-centered normal distribution with precision τ, then priors will be needed for υ, α, β, and τ. We will start by fitting the simpler gap-time model, in which the time scale is reset to zero at the start of each at-risk period. υ describes the shape of the baseline hazard with $\upsilon = 1$ corresponding to the situation in which the risk of an asthma attack does not change with time. We must not be influenced in our choice of prior by the initial exploration of the data and will adopt a gamma distribution with a mean of 1 and not too large a variance, $G(5, 0.2)$, which becomes $G(5, 5)$ in the WinBUGS parameterization.

The scale parameter depends on the unit of time, which in this dataset is days. The chance of an attack on any given day is small; if a child has an asthma attack every 100 days and the hazard is constant over time, then $\exp(\alpha) = 0.01$, so $\alpha = -4.6$. These are high-risk children, so we would be surprised if they went 1,000 days without an attack and so would expect $\alpha < -6.9$; conversely, an attack every 10 days would probably get them withdrawn from the trial, so almost certainly, $\alpha > -2.3$. A reasonable choice seems to be $N(-4.5, 1)$.

The only covariate in this trial is the treatment; if the active treatment halves the hazard, it would be impressive, so we would be surprised if $\beta < -0.7$. On the grounds that for the study to be ethical, our prior should be centered on zero, we might opt for a zero-centered normal distribution with a standard deviation of 0.4, which becomes $N(0, 6.25)$ when parameterized in terms of the precision.

Finally, we have to consider the precision of the frailty; the trial was carried out in a fairly homogeneous group of high-risk children, so we would not expect massive variations in hazard: perhaps some children are at five or even ten times the risk of others but probably not more. So we expect the standard deviation of the random effect (which has to be exponentiated) to be about 0.5, so the precision should be about 4. A $G(2, 2)$ distribution captures this and allows for some uncertainty; this distribution is written as $G(2, 0.5)$ using WinBUGS's parameterization.

Fitting a model to gap time is relatively simple because it becomes a standard Weibull survival model over each at-risk period with observations on the same child linked by the common frailty term and his or her covariates. In this case, the model file is begin: gap model file

```
model {
  for (i in 1:1037) {
     log(lambda[i]) <- alpha + beta*trt[i] + u[id[i]]
     t[i] ~ dweib(v,lambda[i]) C(tcen[i],)
  }
  for (j in 1:232) {
     u[j] ~ dnorm(0,tau)
  }
  alpha ~ dnorm(-4.5,1)
  beta ~ dnorm(0,0.4)
  tau ~ dgamma(2,0.5)
  v ~ dgamma(5,5)
}
```

end: gap model file

The data can be prepared and the model fit using the code

begin: gap in Stata

```
generate t = stop - start
generate tcen = (event==0)*t
replace t = . if event == 0
wbslist (vector id trt t tcen, format(%5.0f)) using data.txt, replace
wbslist (alpha=-4.5,beta=0,tau=4,v=1,u=c(232{0.5*rnormal()})) ///
  using inits.txt, replace
wbsscript using script.txt, replace                         ///
  data(data.txt) model(modelgap.txt) init(inits.txt)        ///
  burnin(2000) update(20000) thin(5) dic logfile(log.txt) ///
  set(alpha beta tau v) coda(duchgap) openbugs
wbsrun using script.txt, openbugs
wbscoda using duchgap, clear openbugs
save duchgapmcmc.dta, replace
mcmcstats alpha beta tau v
```

end: gap model in Stata

The shape parameter, v, shows that the baseline hazard is decreasing, and the treatment effect, `beta`, is similar to the value of -0.254 found by Duchateau et al. (2003) in their likelihood analysis.

```
------------------------------------------------------------------------
Parameter          n      mean        sd       sem   median        95% CrI
------------------------------------------------------------------------
 alpha          4000    -4.301     0.152    0.0064   -4.296 (  -4.609,   -4.013 )
 beta           4000    -0.259     0.117    0.0023   -0.259 (  -0.483,   -0.025 )
 tau            4000     2.445     0.431    0.0092    2.397 (   1.724,    3.422 )
 v              4000     0.848     0.024    0.0011    0.847 (   0.801,    0.897 )
------------------------------------------------------------------------
```

A second issue arises when recurrent events are modeled using calendar time; now the risk periods are left-truncated; that is, each risk period has a start time, and within that risk period, only events that occur after that time are observed. The probability of an event at time t becomes $f(t)/P(t \geq t_L)$, where t_L is the left-truncation time. One of the reasons for the popularity of the Weibull distribution is that it is easy to handle mathematically, and the density, $f(t)$, survival function $P(t \geq t_L)$, and left-truncated density are all easily derived. In our parameterization,

$$f(t) = \upsilon\lambda t^{\upsilon-1}\exp(-\lambda t^{\upsilon})$$
$$P(t \geq t_L) = \exp(-\lambda t_L^{\upsilon})$$
$$f(t)/P(t \geq t_L) = \upsilon\lambda t^{\upsilon-1}\exp\{-\lambda(t^{\upsilon} - t_L^{\upsilon})\}$$

Ideally, it should be possible to specify the left-truncation and right-censoring in an OpenBUGS model file; unfortunately, although such a model will compile, it does not run, and as usual, the error message is unhelpful. Nonetheless, we can still fit the model by specifying the log likelihood directly and using the zeros trick.

begin: model file

```
model {
   for (i in 1:1037) {
      log(lambda[i]) <- alpha + beta*trt[i] + u[id[i]]
      logL[i] <- event[i]*(log(lambda[i])+log(v)+(v-1)*log(t[i]))
                 -lambda[i]*(pow(t[i],v)-pow(tleft[i],v))
      zero[i] ~ dloglik(logL[i])
   }
   for (j in 1:232) {
      u[j] ~ dnorm(0,tau)
   }
   alpha ~ dnorm(-4.5,1)
   beta ~ dnorm(0,0.4)
   tau ~ dgamma(2,0.5)
   v ~ dgamma(5,5)
}
```

end: model file

We have renamed two of the variables in this file to make their names more self-explanatory; thus the event or censoring time `stop` becomes `t`, and the start or left-truncation time becomes `tleft`. We also create a data variable of zeros for use in the zeros trick; otherwise, the Stata code needed to run the analysis is standard.

begin: calendar model in Stata

```
rename stop t
rename start tleft
generate zero = 0
wbslist (vector id trt tleft t event zero, format(%5.0f))  ///
  using data.txt, replace
wbslist (alpha=-4.5,beta=0,tau=4,v=1,u=c(232{0.5*rnormal()})) ///
  using inits.txt, replace
wbsscript using script.txt, replace                        ///
  data(data.txt) model(modelcal.txt) init(inits.txt)       ///
  burnin(1000) update(10000) thin(10) dic logfile(log.txt) ///
  set(alpha beta tau v) coda(duchcal) openbugs
wbsrun using script.txt, openbugs
```

end: calendar model in Stata

The results produced by this model are

```
--------------------------------------------------------------------------
Parameter          n      mean       sd       sem    median         95% CrI
--------------------------------------------------------------------------
 alpha          1000    -5.259    0.238    0.0116   -5.252 (  -5.762,   -4.819 )
 beta           1000    -0.303    0.125    0.0075   -0.302 (  -0.552,   -0.058 )
 tau            1000     1.774    0.258    0.0101    1.751 (   1.340,    2.340 )
 v              1000     1.024    0.034    0.0016    1.023 (   0.961,    1.095 )
--------------------------------------------------------------------------
```

The value of `v` is close to 1, indicating a flat hazard over calendar time as we expected. Duchateau et al. (2003) used maximum likelihood to fit a similar model with gamma frailties and found a treatment effect, `beta`, of -0.300.

Duchateau et al. (2003) went on to consider other models that combine calendar and gap time. For instance, there could be a component of the hazard based on calendar time that is Weibull with one set of parameters and a gap-time effect after the first event that is Weibull with a second set of parameters. In this way, treatment effects could be different for the different components of the hazard, or they could be made the same. If the calendar and gap components of hazard are additive, then a general risk period starting at t_L will have

$$h(t) = \upsilon_c \lambda_c t^{\upsilon_c - 1} + \upsilon_g \lambda_g (t - t_L)^{\upsilon_g - 1}$$

By integration, we can derive the cumulative hazard

$$H(t) = \lambda_c t^{\upsilon_c} + \lambda_g (t - t_L)^{\upsilon_g}$$

This leads directly to the density and survival function needed for the likelihood and the effects of truncation:

$$\begin{aligned} f(t) &= h(t) \exp\{-H(t)\} \\ P(t \geq t_L) &= \exp\{-H(t_L)\} \end{aligned}$$

In this way, Bayesian versions of the Weibull models considered by Duchateau et al. (2003) can be programmed as variations on the code used above for the calendar time model. Finally, we might try alternatives to the Weibull that allow a more general pattern of hazard. For instance, the generalized Weibull suggested by Carrasco, Ortega, and Cordeiro (2008) allows the hazard to follow a u-shaped or bathtub pattern.

11.6 Exercises

1. Bray and Wright (1998) described a Bayesian analysis that combined data from nine surveys of the birth prevalence of Down syndrome. The model used a modified logistic function to capture the pattern of dependence on maternal age. Two of the surveys relied on birth certificates for ascertainment and are therefore strongly suspected of underrecording the true prevalence. However, for each

of those surveys, a smaller calibration study was performed to measure the level of ascertainment. The authors made the raw data available via the journal's website. Analyze those data using your own priors, and then check the model fit. Morris, Mutton, and Alberman (2002) give information useful for setting informative priors.

2. Spiegelhalter (1998) analyzed data on cervical cancer. The numbers of cases each year were obtained from the East Anglian Cancer Registry for the period 1971 to 1995 and are given with information on the size of the corresponding female population. The data are subdivided into four age groups and eight health districts. The author presents a model that allows for a change in 1990 associated with improved screening. As well as the basic model, two additional features were added: an adjustment for overdispersion and hierarchical structure across health districts. The author made the raw data available via the journal's website and presented a BUGS program for the analysis. Download them, reproduce the analysis of the basic model using your own priors, and then add overdispersion and the hierarchical structure. Is there any evidence to justify the increased complexity of the modified models?

12 Writing Stata programs for specific Bayesian analysis

12.1 Introduction

Writing a Bayesian program for general use presents slightly different problems from those encountered when analyzing one's own data. In particular, the programmer has to specify a family of priors without having access to the user's actual beliefs. The aim must be to offer as broad a range of priors as possible and to ensure that some of those priors are noninformative. The programmer's other concern will be to use well-mixing samplers and efficient code so that the program runs as quickly as possible. Almost certainly, this means programming the algorithm in Mata and providing a Stata ado-file that handles the communication with the Mata code.

It is not a good idea to try to control the length of the Markov chain Monte Carlo run or to attempt to assess convergence automatically; this could never be guaranteed to work and would merely give the user a false sense of security. It is much better to create a flexible program and to leave those responsibilities to the user. To illustrate the process of preparing a flexible program for general use, we will write a program that performs a type of regression analysis known as the Bayesian lasso.

12.2 The Bayesian lasso

The least absolute shrinkage and selection operator, known commonly as the lasso, is a technique suggested by Tibshirani (1996) that is closely related to ridge regression. It was developed from a non-Bayesian perspective to address the twin problems of estimate stabilization and variable selection in regression. Rather than using least squares to fit the regression model,

$$y_i = \mu + \sum_{j=1}^{p} b_j x_{ij} + \epsilon_i \quad i = 1, \ldots, n$$

a penalty term is added so that for some user-specified value of λ, the lasso minimizes

$$\sum_{i=1}^{n} (y_i - \mu - \sum_{j=1}^{p} b_j x_{ij})^2 + \lambda \sum_{j=1}^{p} |b_j|$$

The impact of the penalty is to shrink the regression coefficients toward zero. In particular, covariates that add little to the performance of the regression tend to have their coefficients shrunk close to zero, thus making it clearer which explanatory variables are important and which are not. This technique combines variable selection with model fitting and is an alternative to searching through all possible combinations of covariates or to using a stepwise search.

A Bayesian would view this problem slightly differently. The regression model is the same,

$$y_i \sim N(\mu + \sum_{j=1}^{p} b_j x_{ij}, \phi)$$

where ϕ represents the precision, but a Bayesian would set priors on the model parameters rather than apply a penalty. It would be possible, at least in theory, to take each regression coefficient, b_j, and give it an informative prior based on our knowledge of the specific problem. A more realistic alternative would be to imagine a hierarchical structure in which the coefficients are chosen from some common higher-level distribution. This process would make sense only if we first standardize the explanatory variables; otherwise, the coefficients would take values dependent on the units of measurement of the covariates, and it would be hard to imagine such coefficients as coming from a common distribution.

Assuming that the explanatory variables have all been standardized by subtracting their means and dividing by their standard deviations, we can interpret all the coefficients as representing the change in response, y, for a one standard-deviation change in the appropriate covariate. It might then be reasonable to assume that these standardized coefficients were all drawn from a common zero-centered normal distribution. However, because of the shape of the normal distribution, this would encourage lots of moderately sized coefficients and shrink any apparently large coefficients toward the majority. In fact, this assumption would be equivalent to applying a quadratic penalty and would produce a Bayesian version of ridge regression.

When the researcher believes that a few of the explanatory variables will show strong effects but that most will show little or no effect, then it is more appropriate to model the coefficients as coming from a common distribution with a peak at zero and long tails. One option is a Laplace distribution, sometimes called a zero-centered double exponential distribution, which has the form

$$f(b) = \frac{\lambda}{2} \exp(-\lambda|b|)$$

Figure 12.1 shows the shape of this distribution when $\lambda = 2$. With this parameterization, the variance is $2/\lambda^2$, so the larger the value of λ, the larger will be the peak at zero, and the shorter will be the tails. The use of this prior shrinks most coefficients toward zero but leaves a few that are large. Thus it will closely mirror the results obtained from the lasso, and not surprisingly, it has become known as the Bayesian lasso. In fact, if we take the mode of the posterior distribution of each coefficient as its point estimate, then there will be a direct mathematical equivalence between the Bayesian lasso and Tibshirani's (1996) original non-Bayesian method.

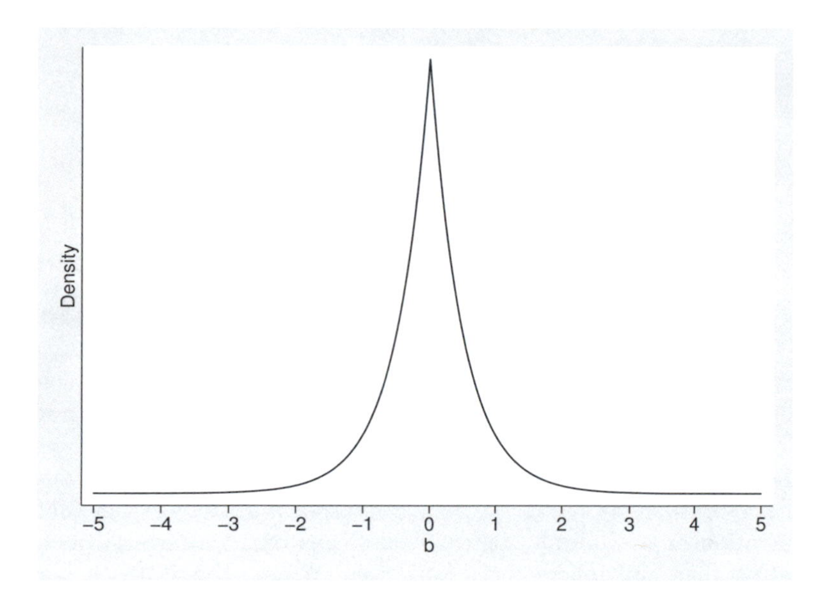

Figure 12.1. Laplace distribution when $\lambda = 2$

In preparing a program for fitting the Bayesian lasso, the programmer must select a family of prior distributions for ϕ, μ, and λ. Because we can hope only to approximate the end-user's actual beliefs, one of the main considerations in choosing the priors is the efficiency of the resulting Markov chain Monte Carlo algorithm. Park and Casella (2008) suggested a neat formulation of this problem that we will follow, except that where they used improper priors for μ and ϕ, we will use normal and gamma distributions, respectively. These replacements not only enable the user to specify vague priors that are effectively equivalent to those used by Park and Casella (2008) but also make it possible to use informative priors when appropriate.

12.3 The Gibbs sampler

The formulation used by Park and Casella (2008) hinges on the fact that the Laplace distribution can be generated through a two-stage process. First, select a random variance from an exponential distribution with mean $2/\lambda^2$. Second, generate a value from a normal distribution with zero mean and that variance. The result of the two stages is a value from a zero-centered Laplace distribution with parameter λ.

With a slight adaptation of the notation of Park and Casella (2008), the full model can be described as

$$\begin{aligned}
y_i &\sim N(\mu + \sum_{j=1}^{p} \beta_j x_{ij}, \phi) \qquad i = 1, \ldots, n \\
\mu &\sim N(m_\mu, t_\mu) \\
\beta_j &\sim N(0, \tau_j \phi) \qquad j = 1, \ldots, p \\
\phi &\sim G(a_\phi, b_\phi) \\
\tau_j^{-1} &\sim G(1, \frac{2}{\omega}) \qquad j = 1, \ldots, p \\
\omega &\sim G(a_\omega, b_\omega)
\end{aligned}$$

where m_μ, t_μ, a_ϕ, b_ϕ, a_ω, and b_ω are numbers specified by users to reflect their own prior beliefs. The parameter of the Laplace distribution, λ, is equal to the square root of ω, so the choice of the prior on ω must reflect the amount of shrinkage that we expect to see. Apart from avoiding improper priors, the major difference between this parameterization and that used by Park and Casella (2008) is the use of precisions rather than variances in the specification of the normal distributions; thus our τ_j are multipliers for those precisions rather than multipliers of the variances. We use the Stata parameterization of the gamma distribution, even though that makes the formulas a little more messy.

This specification is particularly appealing because with a little algebra, all the conditional distributions take recognizable forms and can be used to create efficient Gibbs samplers. The conditional distributions can be identified by first writing the full likelihood times the priors and picking out the terms that contain the parameter of interest. In this case, the likelihood times prior takes the form

$$\prod_{i=1}^{n} p(y_i|\mu, \beta, \phi) \prod_{j=1}^{p} \{p(\beta_j|\tau_j, \phi) p(\tau_j|\omega)\} \, p(\mu) p(\omega) p(\phi)$$

where the distribution of τ_j has to be derived by parameter transformation from that of τ_j^{-1}.

So, for instance, for μ, we would select the terms

$$\exp\left\{-\frac{\phi}{2}\sum_{i=1}^{n}\left(y_i - \mu - \sum_{j=1}^{p}\beta_j x_{ij}\right)^2\right\} \exp\left\{-\frac{t_\mu}{2}(\mu - m_\mu)^2\right\}$$

which after expanding and dropping factors that do not involve μ becomes

$$\exp\left(-\frac{1}{2}\left[\mu^2(n\phi + t_\mu) - 2\mu\left\{\phi\sum_{i=1}^{n}(y_i - \sum_{j=1}^{p}\beta_j x_{ij}) + t_\mu m_\mu\right\}\right]\right)$$

which is recognizable as being equivalent to a normal distribution

$$\mu \sim N \left\{ \frac{\phi \sum_{i=1}^{n} (y_i - \sum_{j=1}^{p} \beta_j x_{ij}) + t_\mu m_\mu}{n\phi + t_\mu}, n\phi + t_\mu \right\}$$

The algebra is rather tedious, but its complexity is one of the main reasons for preparing the program; it makes life much easier for the end user.

Considering the conditional distributions of the other parameters, we find that

$$\beta_j \sim N \left\{ \frac{\sum_{i=1}^{n} x_{ij} (y_i - \mu - \sum_{k \neq j}^{p} \beta_k x_{ik})}{\sum_{i=1}^{n} x_{ij}^2 + \tau_j}, \phi \left(\sum_{i=1}^{n} x_{ij}^2 + \tau_j \right) \right\}$$

$$\phi \sim G \left[\frac{n}{2} + \frac{p}{2} + a_\phi, \left\{ \frac{1}{2} \sum_{i=1}^{n} (y_i - \mu - \sum_{j=1}^{p} \beta_j x_{ij})^2 + \frac{1}{2} \sum_{j=1}^{p} \tau_j \beta_j^2 + \frac{1}{b_\phi} \right\}^{-1} \right]$$

$$\tau_j \sim \text{IGauss} \left(\sqrt{\frac{\omega}{\phi \beta_j^2}}, \omega \right)$$

$$\omega \sim G \left\{ p + a_\omega, \left(\sum_{j=1}^{p} \frac{1}{2\tau_j} + \frac{1}{b_\omega} \right)^{-1} \right\}$$

Thus we can cycle through the parameters, sampling new values directly from these conditional distributions without worrying about whether the simulated values need to be accepted.

Simulating values from the normal and gamma distributions is easy in Stata because of the `rnormal()` and `rgamma()` functions. Unfortunately, there is no function for generating a value from the inverse Gaussian distribution needed for τ_j, so we will need to write our own. An algorithm for generating a value from IGauss(μ, λ) is

- draw $v \sim N(0,1)$;
- calculate $y = v^2$;
- calculate $x = \mu + \frac{\mu^2 y}{2\lambda} - \frac{\mu}{2\lambda} \sqrt{4\mu\lambda y + \mu^2 y^2}$;
- draw $u \sim U(0,1)$; and
- select x if $u \leq \frac{\mu}{\mu + x}$ or select $\frac{\mu^2}{x}$ if $u > \frac{\mu}{\mu + x}$.

Our final consideration before coding the program is to specify the initial values. It is good practice to enable the user to override any choices that we might make because we cannot anticipate every use to which the program will be put. For most users, a good default option will be helpful. We could base our initial values for the coefficients on plausible but not tuned values; for instance, setting all the b_j to 0 and all the τ_j to 1. Or we could modify them to reflect the user's data; for instance, we could set the b_j to be equal to the coefficients from a standard regression.

12.4 The Mata code

First, we need a function to generate a random value from an inverse Gaussian distribution. The following Mata code efficiently creates an $R \times C$ matrix of such values.

begin: inverse Gaussian

```
real matrix RIGAUSS(
   real scalar R,
   real scalar C,
   real scalar mu,
   real scalar lambda)
{
    N = rnormal(R,C,0,1)
    Y = N:*N
    X = mu:+0.5:*mu:*(mu:*Y-sqrt(4:*mu:*lambda:*Y:+mu:*mu:*Y:*Y)):/lambda
    U = runiform(R,C):< mu:/(mu:+X)
    V = U:*X :+ (1:-U):*(mu:*mu:/X)
    return(V)
}
```

end: inverse Gaussian

The Mata program BLASSO implements the Bayesian lasso by Gibbs sampling, assuming that the X's have already been standardized. Most of the code is self-explanatory and merely involves calculating the parameters of the conditional distributions and simulating the updates. The results are written to a comma-delimited text file using the Mata function intokens(), which converts a vector of strings into a string scalar.

begin: Bayesian lasso

```
void BLASSO(
   string scalar Yname,string scalar Xname,real scalar mmu,real scalar tmu,
   real scalar aphi,real scalar bphi,real scalar aomega,real scalar bomega,
   real scalar burnin,real scalar updates,real scalar thin,
   string scalar filename)
{
   Y = st_data(.,Yname)
   X = st_data(.,Xname)
   N = rows(Y)
   P = cols(X)
/*-----------------------
  Initial Values
-----------------------*/
   SSx = colsum(X:*X)
   mu = sum(Y)/N
   phi = 1/(Y´*Y/N-mu*mu)
   B = J(P,1,0)
   T = J(P,1,1)
   omega = 1
   FIT = J(N,1,mu)
```

```
/*-----------------------
  iterations
-----------------------*/
   fh = fopen(filename,"w")
   param = "mu,phi,omega"
   param = param + invtokens(",b":+strofreal(range(1,P,1)´))
   param = param + invtokens(",t":+strofreal(range(1,P,1)´))
   fput(fh,param)
   for (iter=1;iter<=(burnin+updates);iter++) {
   /*-----------------------
    update B
   -----------------------*/
      for (j=1;j<=P;j++) {
         v = B[j]
         prc = T[j] + SSx[j]
         m = X[,j]´*(Y-FIT+v*X[,j])
         B[j] = rnormal(1,1,m/prc,1/sqrt(phi*prc))
         FIT = FIT + (B[j]-v):*X[,j]
      }
   /*-----------------------
    update MU
   -----------------------*/
      v = mu
      m = mmu*tmu + phi*sum(Y-FIT:+v)
      prc = N*phi+tmu
      mu = rnormal(1,1,m/prc,1/sqrt(prc))
      FIT = FIT :+ (mu - v)
   /*-----------------------
    update PHI
   -----------------------*/
      b = 2/bphi + (Y-FIT)´*(Y-FIT) + sum(T:*B:*B)
      a = N/2+P/2+aphi
      phi = rgamma(1,1,a,2/b)
   /*-----------------------
    update OMEGA
   -----------------------*/
      b = 1/bomega + sum(0.5:/T)
      a = P + aomega
      omega = rgamma(1,1,a,1/b)
   /*-----------------------
    update t
   -----------------------*/
      for (j=1;j<=P;j++) T[j] = RIGAUSS(1,1,sqrt(omega/(phi*B[j]*B[j])),omega)
   /*-----------------------
    save estimates
   -----------------------*/
      if(iter > burnin & mod(iter-burnin,thin) == 0) {
         theta = mu,phi,omega,B´,T´
         param = strofreal(mu[1]) +
             invtokens(",":+strofreal(theta[range(2,3+2*P,1)]))
         fput(fh,param)
      }
   }
   fclose(fh)
}
```

end: Bayesian lasso

12.5 A Stata ado-file

A user working in Stata needs a Stata ado-file that communicates with the Mata program. The following code does this:

begin: Stata calling program

```
program bayeslasso
   syntax varlist [if] [in] using/, MU(numlist max=2 min=2)  ///
      PHi(numlist max=2 min=2) OMega(numlist max=2 min=2)    ///
     [Burnin(integer 1000) Updates(integer 10000) Thin(integer 1)]

   preserve
   marksample touse
   quietly drop if `touse´ == 0
*------------------------------
* Split Y and X then standardize
*------------------------------
   local j = 0
   foreach v of varlist `varlist´ {
      if `j´ == 0 local Y "`v´"
      else {
         tempvar s`j´
         egen `s`j´´ = std(`v´)
         local xlist "`xlist´ `s`j´´"
      }
      local ++j
   }
*------------------------------
* Read the prior parameters
*------------------------------
   local mmu : word 1 of `mu´
   local tmu : word 2 of `mu´
   local aphi : word 1 of `phi´
   local bphi : word 2 of `phi´
   local aomega : word 1 of `omega´
   local bomega : word 2 of `omega´
*------------------------------
* Run mata
*------------------------------
   capture rm `using´
   mata: BLASSO("`Y´","`xlist´",`mmu´,`tmu´,`aphi´,`bphi´,`aomega´, ///
     `bomega´,`burnin´,`updates´,`thin´,"`using´")
   restore
end
```

end: Stata calling program

This ado-file works perfectly well but is very basic. A fuller version might include many more data checks and options. Possible additions might include the following:

- user-specified initial values
- checks on the validity of the prior parameters
- displays of the run time

12.6 Testing the code

A vital part of developing any program for general use is its testing, during which one tries to establish both that the program gives the right answers and that it is as robust as possible against the specification of wild starting values or extreme priors. As we noted in chapter 6, two useful ways of establishing that the program works correctly are to use it to analyze simulated data for which the correct answer is known or to use it to analyze a dataset that has been analyzed in the literature using a different algorithm.

Before embarking on the type of validation suggested by Cook, Gelman, and Rubin (2006), we might perform a less formal investigation using simulated data of the type below in which y is generated to depend on x_1 and x_2, but the other variables $x_3, \ldots, x_6$ are independent of y.

begin: simulating data

```
clear
set obs 200
set seed 450892
forvalues j=1/6 {
   generate x`j´ = rnormal()
}
generate y = 5 + x1 + 2*x2 + 2*rnormal()
regress y x1-x6
timer clear 1
timer on 1
bayeslasso y x1-x6 using temp.csv, mu(0 0.0001) phi(0.01 100) omega(50 50)
timer off 1
timer list 1
import delimited temp.csv, clear
generate sd = 1/sqrt(phi)
mcmcstats *
```

end: simulating data

The issues here are whether `bayeslasso` gives coefficients that fit with the simulation and how long the run takes. `mcmcstats` summarizes the results, and the timer commands display the timing. If the program works satisfactorily, then we might investigate whether putting a strong prior on ω gives the anticipated estimates of λ or whether a larger λ produces greater shrinkage. Another variation might be to look at the impact of using highly correlated explanatory variables.

Almost as important as the checks that we make is to remember what we do not need to do. This process is not about establishing that the Bayesian lasso is a good technique, so it is not essential to show that it makes good predictions or that it selects good explanatory variables. There is no need to compare the Bayesian lasso with least-squares regression or stepwise selection unless that helps establish that the code is correct.

12.7 Case study 16: Diabetes data

Efron et al. (2004) wrote an article describing an algorithm called least-angle regression that provides an efficient way of fitting the original lasso of Tibshirani (1996). In that article, they presented a dataset containing 10 potential explanatory variables for a continuous measure of diabetes progression.

To analyze those data using the Bayesian lasso requires us to put priors on μ, ϕ, and ω. The scale of the response variable ranges from just over 0 to about 350; strictly, we should base our prior for μ on our knowledge of the scale rather than the collected data, but a normal distribution with a mean of 175 and a standard deviation of 50 (precision = 0.0004) seems reasonable. The standard deviation of the responses is expected to be under 100, and regression on the explanatory variables should bring that down still further. We might suppose a residual standard deviation of something between 10 and 75 (precision, ϕ, between 0.0002 and 0.01). A gamma distribution such as $G(0.01{,}0.05)$ is reasonable. Setting a prior on λ is extremely difficult because it requires us to specify the amount of shrinkage we expect. A value close to 0 indicates little shrinkage, while a value above 10 results in considerable shrinkage toward 0. We will adopt a realistically vague gamma prior, $G(0.04{,}25)$, which has a mean of 1 and a standard deviation of 5.

Table 12.1 shows the least-squares regression coefficients alongside those given by the Bayesian lasso under our vague priors. The regression coefficients are seen to have been shrunk toward zero with those that were least precisely estimated suffering the largest changes. These Bayesian results were obtained from a burn-in of 1,000 followed by 20,000 updates for which convergence and mixing seemed good. This size of the chain takes about 13 seconds to produce. The fit suggests moderate shrinkage with a posterior mean for λ of 4.1, 95% CrI (1.9,7.1).

Table 12.1. Standardized coefficients from least-squares regression and the Bayesian lasso estimates for the diabetes data

		Least squares		Bayesian lasso	
	parameter	coef	se	mean	sd
	μ	152.1	2.6	152.2	2.6
1	age	−0.5	2.9	−0.2	2.6
2	sex	−11.4	2.9	−10.4	2.9
3	bmi	24.8	3.2	24.9	3.1
4	map	15.5	3.1	14.8	3.1
5	tc	−37.7	19.8	−10.7	10.4
6	ldl	22.7	16.1	1.6	8.7
7	hdl	4.8	10.1	−6.7	6.1
8	tch	8.4	7.7	5.0	6.0
9	ltg	35.8	8.2	25.6	5.3
10	glu	3.2	3.1	3.1	2.9

Figure 12.2 shows the implied prior and posterior distributions of λ.

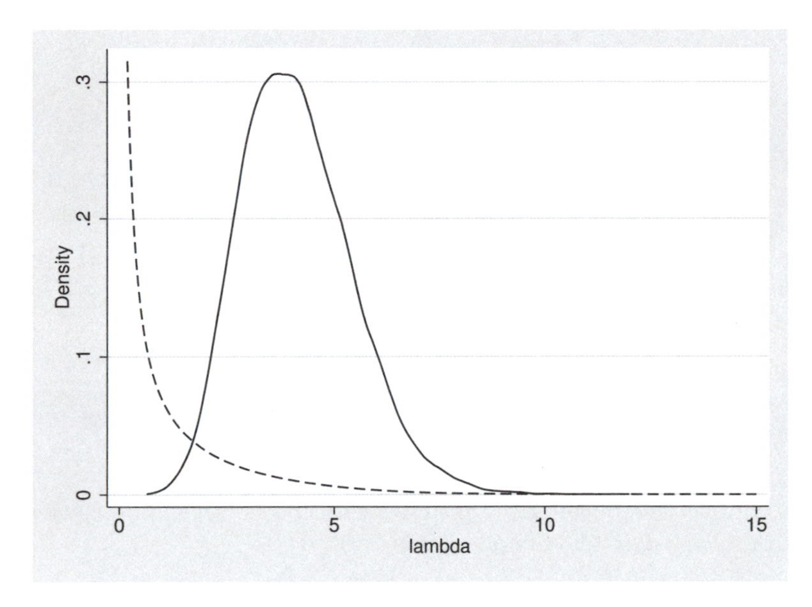

Figure 12.2. Prior (dashed) and posterior (solid) distributions for λ

If we suppress the Gibbs sampler for λ, the value will be fixed, and we can investigate the impact that specific λ's have on the regression coefficients. Perhaps this, too, could be offered as an option with the Stata ado-file. Figure 12.3 shows the changes in the coefficients as λ moves from 0 to 10. Increasing λ further causes a slow but continual shrinkage toward 0 until by $\lambda = 1{,}000$, the estimated coefficients are all within 2 posterior standard deviations of 0.

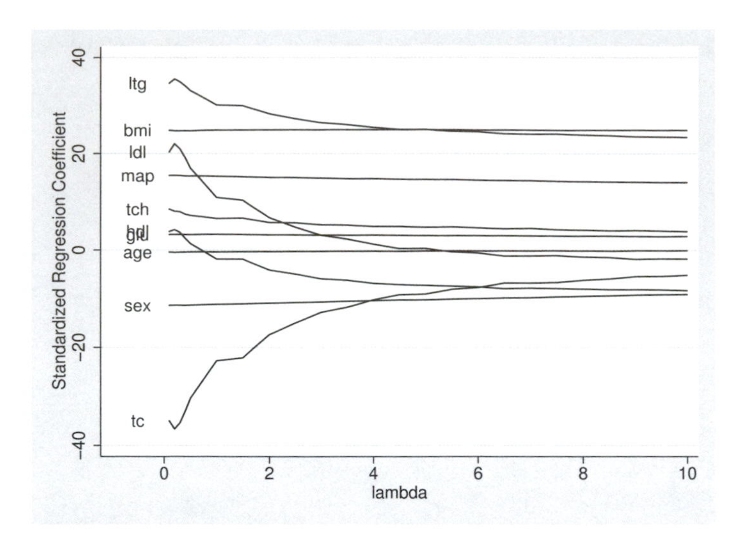

Figure 12.3. Impact that changing λ has on the estimated regression coefficients; numbers refer to the covariate identifiers in table 12.1

Figure 12.4 shows the estimates of the coefficients and 95% credible intervals (CrI) when $\lambda = 10$ and the least-squares estimates. At this degree of shrinkage, we can be confident only in the effects of four explanatory variables, `sex`, `bmi`, `map`, and `ltg`, while the analysis without shrinkage suggests that `tc` and `ldl` also have large estimated effects but wide uncertainty.

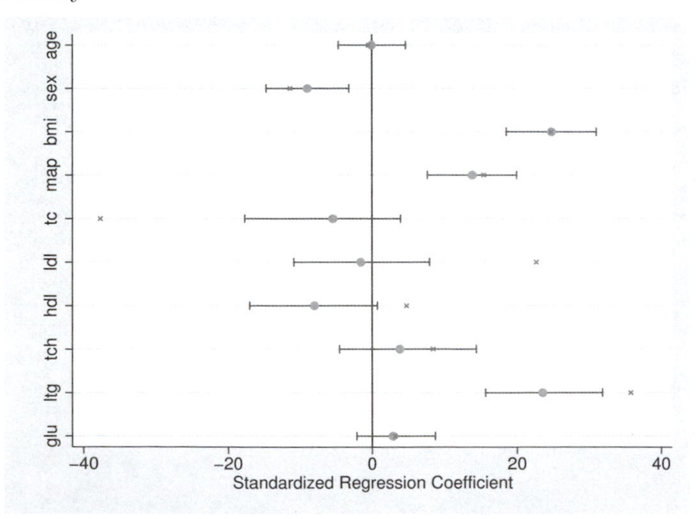

Figure 12.4. Estimated regression coefficients with $\lambda = 10$; crosses show the corresponding least-squares estimates

12.8 Extensions to the Bayesian lasso program

Following Park and Casella (2008), the hierarchical structure used for the regression coefficients took the form

$$\beta_j \sim N(0, \tau_j \phi) \qquad j = 1, \ldots, p$$
$$\tau_j^{-1} \sim G\left(1, \frac{2}{\omega}\right) \qquad j = 1, \ldots, p$$

That is, the coefficients are drawn from normal distributions with precisions that are multiples of the residual precision, ϕ. This implies a common Laplace distribution with the form

$$p(\beta_j) = \frac{\lambda\sqrt{\phi}}{2} \exp\left(-\lambda\sqrt{\phi}|b_j|\right)$$

so that the shape of the Laplace distribution varies with both $\lambda = \sqrt{\omega}$ and ϕ. A more natural choice might have been to use

$$p(\beta_j) = \frac{\lambda}{2} \exp\left(-\lambda|b_j|\right)$$

because with this distribution, it is easier to place a prior on λ. Park and Casella (2008) also consider this distribution but caution against it on the grounds that it can lead to bimodal posterior distributions for the b_j, making convergence slower and interpretation more difficult.

Park and Casella (2008) also discuss Bridge regression as first suggested by Frank and Friedman (1993), the basic form of which is based on minimizing

$$\sum_{i=1}^{n}(y_i - \mu - \sum_{j=1}^{p} b_j x_{ij})^2 + \lambda \sum_{j=1}^{p} |b_j|^q \qquad q \geq 0$$

The lasso ($q = 1$) and ridge regression ($q = 2$) are special cases of this general form. Kyung et al. (2010) discuss Bayesian algorithms for a modification of the lasso known as the fused lasso, which is equivalent to minimizing

$$\sum_{i=1}^{n}(y_i - \mu - \sum_{j=1}^{p} b_j x_{ij})^2 + \lambda_1 \sum_{j=1}^{p} |b_j| + \lambda_2 \sum_{j=2}^{p} |b_j - b_{j-1}|$$

As the formula suggests, this can be used only when there is a natural ordering of the explanatory variables. They also consider the grouped lasso for situations in which the explanatory variables have a natural grouping and the elastic net that includes two penalty terms: one involving the sum of the absolute values of the coefficients and the other the sum of their squares.

The basic approach underlying the lasso could also be applied to logistic regression, Poisson regression, or any other form of regression. It would also be easy to define

Bayesian versions of these models. However, to create an efficient, generally usable program requires a Gibbs sampler or other fast algorithm. This is a topic of current research, and at present, such algorithms are not always available.

12.9 Exercises

1. Modify the program for the Bayesian lasso given in this chapter to use the hierarchical distribution

$$p(\beta_j) = \frac{\lambda}{2} \exp\left(-\lambda |b_j|\right)$$

in place of

$$p(\beta_j) = \frac{\lambda\sqrt{\phi}}{2} \exp\left(-\lambda\sqrt{\phi}|b_j|\right)$$

and use it to fit a regression model to the diabetes data. Is there any evidence of bimodal posterior distributions for this dataset?

2. Ridge regression minimizes

$$\sum_{i=1}^{n}(y_i - \mu - \sum_{j=1}^{p} b_j x_{ij})^2 + \lambda \sum_{j=1}^{p} b_j^2$$

and a Bayesian version of this method can be obtained by sampling the coefficients, b_j, from a common normal distribution rather than the Laplace distribution used in the lasso. Prepare a generally usable program for Bayesian ridge regression.

A Standard distributions

Bernoulli

Special case of the binomial with $n = 1$.

$$p(y;p) = p^y(1-p)^{1-y} \qquad y = 0,1 \quad 0 \le p \le 1$$

mean $= p$, variance $= p(1-p)$

Stata: `logdensity bernoulli logp y p`
Mata: `logdensity("bernoulli",logp,y,p)`
WinBUGS: `y ~ dbern(p)`

Beta

Often used as the prior for the proportion in a binomial or Bernoulli model. Beta(1,1) is the same as a uniform(0,1) distribution.

$$p(y;\alpha,\beta) = \frac{\Gamma(\alpha+\beta)}{\Gamma(\alpha)\Gamma(\beta)} y^{\alpha-1}(1-y)^{\beta-1} \qquad 0 \le y \le 1 \quad 0 < \alpha,\beta$$

mean $= \frac{\alpha}{\alpha+\beta}$, variance $= \frac{\alpha\beta}{(\alpha+\beta)^2(\alpha+\beta+1)}$

For a prior with mean, m, and variance, v, use

$$\alpha = m\left\{\frac{m(1-m)}{v} - 1\right\} \text{ and } \beta = (1-m)\left\{\frac{m(1-m)}{v} - 1\right\}$$

Stata: `logdensity beta logp y alpha beta`
Mata: `logdensity("beta",logp,y,alpha,beta)`
WinBUGS: `y ~ dbeta(alpha,beta)`

Binomial

Used as a model for the number of occurrences in n independent trials when the probability of the event is p.

$$p(y; p, n) = \frac{n!}{y!(n-y)!} p^y (1-p)^{n-y} \qquad y = 0, 1, \ldots, n \qquad 0 \leq p \leq 1$$

mean $= np$, variance $= np(1-p)$

Stata: `logdensity binomial logp y n p` omits the factorial terms
Mata: `logdensity("binomial",logp,y,n,p)` omits the factorial terms
WinBUGS: `y ~ dbin(p,n)`

Categorical

Used for a random selection from n coded outcomes, $y = 1, 2, \ldots, n$, with probabilities, p_y.

$$p(y) = p_y \qquad y = 1, 2, \ldots, n \qquad \sum_{y=1}^{n} p_y = 1$$

mean $\mu = \sum_{y=1}^{n} p_y y$, variance $= \sum_{y=1}^{n} p_y (y-\mu)^2$

Stata: `logdensity categorical logp y p`
Mata: `logdensity("categorical",logp,y,p)`
WinBUGS: `y ~ dcat(p[])`

Chi-squared

Rarely used in Bayesian models. The chi-squared distribution with k degrees of freedom is a special case of the gamma distribution with $\alpha = k/2$ and $\beta = 2$. k is usually but not necessarily an integer.

$$p(y;k) = \frac{y^{k/2-1}\exp(-y/2)}{2^{k/2}\Gamma(k/2)} \qquad 0 < y \quad 0 < k$$

mean $= k$, variance $= 2k$

Stata: `logdensity chisqr logp y k`
Mata: `logdensity("chisq",logp,y,k)`
WinBUGS: `y ~ dchisqr(k)`

Dirichlet

Often used as the prior for a set of probabilities that sum to 1. Conjugate to the multinomial distribution.

$$p(y[\,];\alpha[\,]) = \frac{\Gamma(\sum_{i=1}^n \alpha_i)}{\prod_{i=1}^n \Gamma(\alpha_i)} \prod_{i=1}^n y_i^{\alpha_i - 1} \qquad 0 < y_i < 1 \quad \sum_{i=1}^n y_i = 1 \quad 0 < \alpha_i$$

mean$(y_i) = \frac{\alpha_i}{\alpha_0}$, variance$(y_i) = \frac{\alpha_i(\alpha_0 - \alpha_i)}{\alpha_0^2(\alpha_0+1)}$, covariance$(y_i, y_j) = \frac{\alpha_i \alpha_j}{\alpha_0^2(\alpha_0+1)}$

where $\alpha_0 = \sum_{i=1}^n \alpha_i$.

Stata: `logdensity dirichlet logp y alpha`
Mata: `logdensity("dirichlet",logp,y,alpha)`
WinBUGS: `y ~ ddirch(alpha[])` OpenBUGS: `y ~ ddirich(alpha[])`

Exponential

Used as a model for times between events in a random process with a constant rate and no memory. Special case of the gamma distribution in which $\alpha = 1$ and $\beta = \mu$.

$$p(y;\mu) = \frac{\exp(-y/\mu)}{\mu} \qquad 0 < y, \quad 0 < \mu$$

mean $= \mu$, variance $= \mu^2$

Winbugs parameterizes the exponential distribution in terms of the rate $\lambda = 1/\mu$.

Stata: `logdensity exponential logp y mu`
Mata: `logdensity("exponential",logp,y,mu)`
WinBUGS: `y ~ dexp(lambda)`

Gamma

Sometimes used as a model for nonnegative observations such as the time to an event. Conjugate prior for the precision in a normal model; also often used as the prior for other nonnegative parameters.

$$p(y;\alpha,\beta) = \frac{y^{\alpha-1}\exp(-y/\beta)}{\beta^{\alpha}\Gamma(\alpha)} \qquad 0 < y \quad 0 < \alpha, \beta$$

α is the shape parameter, β is the scale parameter. $\Gamma(\alpha) = \int_0^\infty t^{\alpha-1}e^{-t}dt$, or $\Gamma(\alpha) = (\alpha-1)!$ if α is an integer.

mean $= \alpha\beta$, variance $= \alpha\beta^2$

For a prior with mean, m, and variance, v, use $\alpha = \frac{m^2}{v}$ and $\beta = \frac{v}{m}$.

WinBUGS parameterizes in terms of α and the rate parameter $\lambda = 1/\beta$.

Stata: `logdensity gamma logp y alpha beta`
Mata: `logdensity("gamma",logp,y,alpha,beta)`
WinBUGS: `y ~ dgamma(alpha,lambda)`

Generalized gamma

Used as a flexible model for survival data that allows nonmonotonic hazards.

$$p(y;\alpha,\beta,\gamma) = \frac{\gamma y^{\alpha-1}\exp(-[y/\beta]^{\gamma})}{\beta^{\alpha}\Gamma(\frac{\alpha}{\gamma})} \qquad 0 < y \quad 0 < \alpha,\beta,\gamma$$

mean $= \frac{\beta\Gamma\{(\alpha+1)/\gamma\}}{\Gamma(\alpha/\gamma)}$, variance $= \beta^2\frac{\Gamma\{(\alpha+2)/\gamma\}\Gamma(\alpha/\gamma)-\Gamma\{(\alpha+1)/\gamma\}^2}{\Gamma(\alpha/\gamma)^2}$

Reduces to the gamma distribution when $\gamma = 1$ and to the Weibull when $\gamma = \alpha$.

WinBUGS parameterizes in terms of γ, $\mu = 1/\beta$, and $r = \alpha/\gamma$.

Stata: `logdensity gengamma logp y alpha beta gamma`
Mata: `logdensity("gengamma",logp,y,alpha,beta,gamma)`
WinBUGS: `y ~ dgen.gamma(r,mu,gamma)`

Inverse gamma

Sometimes used as the prior for the variance in a normal model. Equivalent to using a gamma prior for the precision.

$$p(y;\alpha,\beta) = \frac{\beta^{\alpha}y^{-\alpha-1}}{\Gamma(\alpha)}\exp(-\beta/y) \qquad 0 < y \quad 0 < \alpha,\beta$$

mean $= \beta/(\alpha-1)$ if $\alpha > 1$, variance $= \beta^2/\{(\alpha-1)^2(\alpha-2)\}$ if $\alpha > 2$

WinBUGS does not offer the inverse gamma; however, if $y \sim G(\alpha,\beta)$, then $1/y \sim \text{IG}(\alpha, 1/\beta)$.

The inverse chi-squared distribution with k degrees of freedom is a special case of the inverse gamma distribution in which $\alpha = k/2$ and $\beta = 1/2$.

Stata: `logdensity igamma logp y alpha beta`
Mata: `logdensity("igamma",logp,y,alpha,beta)`
WinBUGS does not have a function for the inverse gamma.

Inverse Gaussian

Used to model survival times or other positive values with a long-tailed distribution. Despite the name, it is not closely related to the normal (Gaussian) distribution.

$$p(y;\lambda,\mu) = \left(\frac{\lambda}{2\pi y^3}\right)^{1/2} \exp\left\{\frac{-\lambda(y-\mu)^2}{2\mu^2 y}\right\} \qquad y>0 \quad 0<\lambda,\mu$$

mean $= \mu$, variance $= \mu^3/\lambda$

Stata: `logdensity igauss logp y mu lambda`
Mata: `logdensity("igauss",logp,y,mu,lambda)`
WinBUGS does not have a function for the inverse Gaussian.

Laplace

Also known as the double exponential distribution because it resembles an exponential distribution reflected in $y = \mu$.

$$p(y;\mu,b) = \frac{1}{2\phi}\exp\left(\frac{-|y-\mu|}{\phi}\right) \qquad -\infty<y<\infty \quad 0<\phi$$

mean $= \mu$, variance $= 2\phi^2$

WinBUGS parameterizes the distribution in terms of μ and $\lambda = 1/\phi$.

Stata: `logdensity laplace logp y mu phi`
Mata: `logdensity("laplace",logp,y,mu,phi)`
WinBUGS: `y ~ ddexp(mu,lambda)`

Logistic

Similar in shape to a normal distribution but with heavier tails.

$$p(y;\mu,\sigma) = \frac{\exp\left(-\frac{y-\mu}{\sigma}\right)}{\sigma\left\{1+\exp\left(-\frac{y-\mu}{\sigma}\right)\right\}^2} \qquad -\infty < y, \mu < \infty, \quad \sigma > 0$$

mean $= \mu$, variance $= \sigma^2\pi^2/3$

WinBUGS parameterizes the logistic in terms of μ and $\tau = 1/\sigma$.

Stata: `logdensity logistic logp y mu sigma`
Mata: `logdensity("logistic",logp,mu,sigma)`
WinBUGS: `y ~ dlogis(mu,tau)`

Log normal

If $\log(y)$ follows a normal distribution, then y follows a log-normal distribution.

$$p(y;\mu,\sigma) = \frac{1}{y\sigma\sqrt{2\pi}} \exp\left\{-\frac{\{\log(y)-\mu\}^2}{2\sigma^2}\right\} \qquad y > 0, \quad -\infty < \mu < \infty \quad 0 < \sigma$$

mean $= \exp(\mu + \sigma^2/2)$, variance $= \exp(2\mu + \sigma^2)\{\exp(\sigma^2) - 1\}$

WinBUGS parameterizes in terms of μ and $\tau = 1/\sigma^2$.

Stata: `logdensity lognormal logp y mu sigma`
Mata: `logdensity("lognormal",logp,y,mu,sigma)`
WinBUGS: `y ~ dlnorm(mu,tau)`

Multinomial

Multivariate discrete generalization of the binomial. Number of responses in each category when there are n independent trials, each of which gives an outcome in one of k categories and the probabilities are p_i, $i = 1, \ldots, k$.

$$p(y; p, n) = \frac{n!}{\prod_i y_i!} \prod_i p_i^{y_i} \qquad \sum_i y_i = n \qquad \sum p_i = 1 \qquad 0 < p_i < 1$$

where $y_i = 0, 1, \ldots, n \quad i = 1, \ldots, k$.

$\text{mean}(y_i) = np_i$, $\text{variance}(y_i) = np_i(1 - p_i)$,

$\text{correlation}(y_i, y_j) = -\sqrt{p_i p_j / (1 - p_i)(1 - p_j)}$

Stata: `logdensity multinomial logp y p` omits the factorial terms
Mata: `logdensity("multinomial",logp,y,p)` omits the factorial terms
WinBUGS: `y[] ~ dmulti(p[],n)`

Multivariate normal

p-dimensional generalization of the normal distribution,

$$p(y; \mu, \Sigma) = \frac{1}{(2\pi)^{p/2} |\Sigma|^{1/2}} \exp\left\{ -\frac{1}{2}(y - \mu)' \Sigma^{-1} (y - \mu) \right\}$$

where y and μ are $p \times 1$ column vectors, and Σ is a $p \times p$ positive definite covariance matrix.

mean vector $= \mu$, variance matrix $= \Sigma$

WinBUGS parameterizes in terms of μ and the $T = \Sigma^{-1}$.

Stata: `logdensity mnormal logp y mu Sigma`
Mata: `logdensity("mnormal",logp,y,mu,Sigma)`
WinBUGS: `y[] ~ dmnorm(mu[],T[,])`

Multivariate Student's t

p-dimensional generalization of the t distribution with k degrees of freedom.

$$p(y;\mu,\Sigma) = \frac{\Gamma\{(k+p)/2\}}{\Gamma(k/2)(k\pi)^{p/2}|\Sigma|^{1/2}} \left\{1+(y-\mu)'\Sigma^{-1}(y-\mu)/k\right\}^{-(k+p)/2}$$

mean vector $= \mu$, variance matrix $= k\Sigma/(k-2)$ if $k > 2$

WinBUGS parameterizes in terms of μ, k, and $T = \Sigma^{-1}$.

Stata: `logdensity mt logp y mu Sigma k`
Mata: `logdensity("mt",logp,y,mu,Sigma,k)`
WinBUGS: `y[] ~ dmt(mu[],T[,],k)`

Negative binomial

The probability of y failures before the kth success. Alternative to the Poisson when the variance is greater than the mean. It can be derived as a mixture of Poisson distributions in which the means are drawn from a gamma distribution. The geometric distribution is a special case of the negative binomial with $k = 1$.

$$p(y;p,k) = \frac{(y+k-1)!}{y!(k-1)!} p^k (1-p)^y \qquad y = 0, 1, \ldots \qquad 1 \leq r \qquad 0 < p < 1$$

mean $= k(1-p)/p$, variance $= k(1-p)/p^2$

Stata and WinBUGS use this parameterization, but occasionally, the negative binomial is written in terms of k and $q = 1 - p$.

Stata: `logdensity negbinomial logp y p k`
Mata: `logdensity("negbinomial",logp,y,p,k)`
WinBUGS: `y ~ dnegbin(p,k)`

Normal

Standard model for measurements that vary symmetrically about their mean and for symmetric random effects. Often used as a prior for parameters that can take any real value.

$$p(y;\mu,\sigma) = \frac{1}{\sigma\sqrt{2\pi}} \exp\left\{-\frac{1}{2\sigma^2}(y-\mu)^2\right\} \qquad -\infty < y < \infty \quad -\infty < \mu < \infty \quad 0 < \sigma$$

mean $= \mu$, variance $= \sigma^2$

WinBUGS parameterizes in terms of μ and the precision $\tau = 1/\sigma^2$.

Stata: `logdensity normal logp y mu sigma`
Mata: `logdensity("normal",logp,y,mu,sigma)`
WinBUGS: `y ~ dnorm(mu,tau)`

Pareto

Long-tailed model for values that cannot be less than c. Primarily used in economics.

$$p(y;a,c) = ac^a y^{-(a+1)} \qquad c \le y \quad 0 < a$$

mean $= ac/(a-1)$ if $\alpha > 1$, variance $= \frac{ac^2}{(a-1)^2(a-2)}$ if $\alpha > 2$

Stata: `logdensity pareto logp y a c`
Mata: `logdensity("pareto",logp,y,a,c)`
WinBUGS: `y ~ dpar(a,c)`

Poisson

Used as a model for counts of events that are randomly spread in time or space with a constant rate.

$$p(y;\lambda) = \exp(-\lambda)\frac{\lambda^y}{y!} \qquad y = 0, 1, \ldots \qquad 0 < \lambda$$

mean $= \lambda$, variance $= \lambda$

Stata: `logdensity poisson logp y lambda`
Mata: `logdensity("poisson",logp,y,lambda)`
WinBUGS: `y ~ dpois(lambda)`

Student's t

Similar to a normal distribution but with wider tails. Sometimes used in place of the normal to give robustness to outliers.

$$p(y;\mu,\sigma,k) = \frac{\Gamma\left(\frac{k+1}{k}\right)}{\sigma\sqrt{k\pi}\Gamma\left(\frac{k}{2}\right)}\left\{1 + \frac{(y-\mu)^2}{k\sigma^2}\right\}^{-\frac{k+1}{2}} \qquad 0 < a$$

mean $= \mu$, variance $= \sigma^2\frac{k}{k-2}$

WinBUGS parameterizes in terms of μ and $\tau = 1/\sigma^2$.

Stata: `logdensity t logp y mu sigma k`
Mata: `logdensity("t",logp,y,mu,sigma,k)`
WinBUGS: `y ~ dt(mu,tau,k)`

Uniform

Often used to create a noninformative (flat) prior.

$$p(y;\alpha,\beta) = \frac{1}{\beta-\alpha} \quad \alpha < y < \beta$$

mean $= (\alpha+\beta)/2$, variance $= (\beta-\alpha)^2/12$

Stata: `logdensity uniform logp y alpha beta`
Mata: `logdensity("uniform",logp,y,alpha,beta)`
WinBUGS: `y ~ dunif(alpha,beta)`

Weibull

Often used to model the time to an event, especially in survival analysis. Reduces to the exponential when $\upsilon = 1$.

$$p(y;\upsilon,\mu) = \frac{\upsilon(y/\mu)^{\upsilon-1}}{\mu}\exp\{-(y/\mu)^{\upsilon}\} \qquad y > 0, \qquad 0 < \upsilon, \mu$$

mean $= \mu\Gamma\left(\frac{\upsilon+1}{\upsilon}\right)$, variance $= \mu^2\left\{\Gamma\left(\frac{\upsilon+2}{\upsilon}\right) - \Gamma\left(\frac{\upsilon+1}{\upsilon}\right)^2\right\}$

Stata and WinBUGS both parameterize in terms of υ and $\lambda = \left(\frac{1}{\mu}\right)^{\upsilon}$.

Stata: `logdensity weibull logp y upsilon mu`
Mata: `logdensity("weibull",logp,y,upsilon,mu)`
WinBUGS: `y ~ dweib(upsilon,lambda)`

Wishart

Commonly used as the conjugate prior for the inverse of the $p \times p$ covariance matrix in a multivariate normal model

$$p(Y;S,k) = \frac{|Y|^{(k-p-1)/2}}{2^{kp/2}\Gamma(k/2)|S|^{k/2}} \exp\left\{-\frac{1}{2}\mathrm{Tr}(S^{-1}Y)\right\} \qquad p \leq k$$

where Y is the random $p \times p$ matrix, S is a $p \times p$ matrix, and k is a scalar.

mean $= kS$

WinBUGS parameterizes in terms of k and $R = S^{-1}$.

Stata: `logdensity wishart logp y S k`
Mata: `logdensity("wishart",logp,y,S,k)`
WinBUGS: `y[,] ~ dwish(R[,],k)`

References

Alund, M., C. Hoe-Hansen, B. Tillander, B. A. Hedén, and R. Norlin. 2000. Outcome after cup hemiarthroplasty in the rheumatoid shoulder: A retrospective evaluation of 39 patients followed for 2–6 years. *Acta Orthopaedica Scandinavica* 71: 180–184.

Baldi, P., and A. D. Long. 2001. A Bayesian framework for the analysis of microarray expression data: Regularized t-test and statistical inferences of gene changes. *Bioinformatics* 17: 509–519.

Baum, C. F. 2009. *An Introduction to Stata Programming.* College Station, TX: Stata Press.

Bayarri, M. J., and J. O. Berger. 1997. Measures of surprise in Bayesian analysis. Working paper 97-46, Institute of Statistics and Decision Sciences, Duke University. http://ftp.isds.duke.edu/WorkingPapers/97-46.ps.

Bayarri, M. J., and M. E. Castellanos. 2007. Bayesian checking of the second levels of hierarchical models. *Statistical Science* 22: 322–343.

Beitler, P. J., and J. R. Landis. 1985. A mixed-effects model for categorical data. *Biometrics* 41: 991–1000.

Berger, J. O. 2006. The case for objective Bayesian analysis. *Bayesian Analysis* 1: 385–402.

Birck, R., S. Krzossok, F. Markowetz, P. Schnülle, F. J. van der Woude, and C. Braun. 2003. Acetylcysteine for prevention of contrast nephropathy: Meta-analysis. *Lancet* 362: 598–603.

Box, G. E. P. 1976. Science and statistics. *Journal of the American Statistical Association* 71: 791–799.

———. 1980. Sampling and Bayes' inference in scientific modelling and robustness. *Journal of the Royal Statistical Society, Series A* 143: 383–430.

Bray, I. 2002. Application of Markov chain Monte Carlo methods to projecting cancer incidence and mortality. *Journal of the Royal Statistical Society, Series C* 51: 151–164.

Bray, I., and D. E. Wright. 1998. Application of Markov chain Monte Carlo methods to modelling birth prevalence of Down syndrome. *Journal of the Royal Statistical Society, Series C* 47: 589–602.

Breslow, N. E. 1984. Extra-Poisson variation in log-linear models. *Journal of the Royal Statistical Society, Series C* 33: 38–44.

Brooks, S. P. 1998. Markov chain Monte Carlo method and its application. *Journal of the Royal Statistical Society, Series D* 47: 69–100.

Brooks, S. P., and A. Gelman. 1998. General methods for monitoring convergence of iterative simulations. *Journal of Computational and Graphical Statistics* 7: 434–455.

Brooks, S. P., and G. O. Roberts. 1998. Convergence assessment techniques for Markov chain Monte Carlo. *Statistics and Computing* 8: 319–335.

Brophy, J. M., L. Joseph, and J. L. Rouleau. 2001. Beta-blockers in congestive heart failure: A Bayesian meta-analysis. *Annals of Internal Medicine* 134: 550–560.

Carlin, B. P., and S. Chib. 1995. Bayesian model choice via Markov chain Monte Carlo methods. *Journal of the Royal Statistical Society, Series B* 57: 473–484.

Carrasco, J. M. F., E. M. M. Ortega, and G. M. Cordeiro. 2008. A generalized modified Weibull distribution for lifetime modeling. *Computational Statistics & Data Analysis* 53: 450–462.

Casella, G., and E. I. George. 1992. Explaining the Gibbs sampler. *American Statistician* 46: 167–174.

Chen, M.-H., and Q.-M. Shao. 1999. Monte Carlo estimation of Bayesian credible and HPD intervals. *Journal of Computational and Graphical Statistics* 8: 69–92.

Chib, S., and E. Greenberg. 1995. Understanding the Metropolis–Hastings algorithm. *American Statistician* 49: 327–335.

Cook, S. R., A. Gelman, and D. B. Rubin. 2006. Validation of software for Bayesian models using posterior quantiles. *Journal of Computational and Graphical Statistics* 15: 675–692.

Cowles, M. K., and B. P. Carlin. 1996. Markov chain Monte Carlo convergence diagnostics: A comparative review. *Journal of the American Statistical Association* 91: 883–904.

Crowder, M. J. 1978. Beta-binomial anova for proportions. *Journal of the Royal Statistical Society, Series C* 27: 34–37.

Dellaportas, P., J. J. Forster, and I. Ntzoufras. 2002. On Bayesian model and variable selection using MCMC. *Statistics and Computing* 12: 27–36.

Dellaportas, P., and A. F. M. Smith. 1993. Bayesian inference for generalized linear and proportional hazards models via Gibbs sampling. *Journal of the Royal Statistical Society, Series C* 42: 443–459.

Delmar, P., S. Robin, D. Tronik-Le Roux, and J. J. Daudin. 2005. Mixture model on the variance for the differential analysis of gene expression data. *Journal of the Royal Statistical Society, Series C* 54: 31–50.

Detrano, R., A. Janosi, W. Steinbrunn, M. Pfisterer, J.-J. Schmid, S. Sandhu, K. H. Guppy, S. Lee, and V. Froelicher. 1989. International application of a new probability algorithm for the diagnosis of coronary artery disease. *American Journal of Cardiology* 64: 304–310.

Draper, E. S., J. Zeitlin, D. J. Field, B. N. Manktelow, and P. Truffert. 2007. Mortality patterns among very preterm babies: A comparative analysis of two European regions in France and England. *Archives of Disease in Childhood: Fetal & Neonatal* 92: F356–F360.

Duchateau, L., P. Janssen, I. Kezic, and C. Fortpied. 2003. Evolution of recurrent asthma event rate over time in frailty models. *Journal of the Royal Statistical Society, Series C* 52: 355–363.

Efron, B., T. Hastie, I. Johnstone, and R. Tibshirani. 2004. Least angle regression. *Annals of Statistics* 32: 407–499.

El Adlouni, S., A.-C. Favre, and B. Bobée. 2006. Comparison of methodologies to assess the convergence of Markov chain Monte Carlo methods. *Computational Statistics & Data Analysis* 50: 2685–2701.

Elston, R. C., and J. E. Grizzle. 1962. Estimation of time-response curves and their confidence bands. *Biometrics* 18: 148–159.

Field, D. J., E. S. Draper, A. Fenton, E. Papiernik, J. Zeitlin, B. Blondel, M. Cuttini, R. F. Maier, T. Weber, M. Carrapato, L. Kollée, J. Gadzin, and P. Van Reempts. 2009. Rates of very preterm birth in Europe and neonatal mortality rates. *Archives of Disease in Childhood: Fetal & Neonatal* 94: F253–F256.

Frank, I. E., and J. H. Friedman. 1993. A statistical view of some chemometrics regression tools. *Technometrics* 35: 109–135.

Friedman, J. H., and J. W. Tukey. 1974. A projection pursuit algorithm for exploratory data analysis. *IEEE Transactions on Computers* C-23: 881–890.

Gelman, A. 2004. Exploratory data analysis for complex models. *Journal of Computational and Graphical Statistics* 13: 755–779.

———. 2006. Prior distributions for variance parameters in hierarchical models (comment on article by Browne and Draper). *Bayesian Analysis* 1: 515–534.

———. 2008. Objections to Bayesian statistics. *Bayesian Analysis* 3: 445–449.

Gelman, A., J. B. Carlin, H. S. Stern, and D. B. Rubin. 2004. *Bayesian Data Analysis*. 2nd ed. Boca Raton, FL: Chapman & Hall/CRC.

Gelman, A., I. V. Mechelen, G. Verbeke, D. F. Heitjan, and M. Meulders. 2005. Multiple imputation for model checking: Completed-data plots with missing and latent data. *Biometrics* 61: 74–85.

Gelman, A., and D. B. Rubin. 1992. Inference from iterative simulation using multiple sequences. *Statistical Science* 7: 457–472.

Geweke, J. 1992. Evaluating the accuracy of sampling-based approaches to the calculation of posterior moments. In *Bayesian Statistics 4*, ed. J. M. Bernado, J. O. Berger, A. P. Dawid, and A. F. M. Smith, 169–193. Oxford: Oxford University Press.

———. 2004. Getting it right: Joint distribution tests of posterior simulators. *Journal of the American Statistical Association* 99: 799–804.

Gilks, W. R., N. G. Best, and K. K. C. Tan. 1995. Adaptive rejection Metropolis sampling within Gibbs sampling. *Journal of the Royal Statistical Society, Series C* 44: 455–472.

Gilks, W. R., R. M. Neal, N. G. Best, and K. K. C. Tan. 1997. Corrigendum: Adaptive rejection Metropolis sampling. *Journal of the Royal Statistical Society, Series C* 46: 541–542.

Gilks, W. R., S. Richardson, and D. J. Spiegelhalter. 1996. *Markov Chain Monte Carlo in Practice*. Boca Raton, FL: Chapman & Hall.

Gilks, W. R., and P. Wild. 1992. Adaptive rejection sampling for Gibbs sampling. *Journal of the Royal Statistical Society, Series C* 41: 337–348.

Goldstein, M. 2006. Subjective Bayesian analysis: Principles and practice. *Bayesian Analysis* 1: 403–420.

Goodman, S. N. 1999. Toward evidence-based medical statistics. 2: The Bayes factor. *Annals of Internal Medicine* 130: 1005–1013.

Gould, W. 2006. Mata matters: Creating new variables—sounds boring, isn't. *Stata Journal* 6: 112–123.

———. 2007. Mata matters: Subscripting. *Stata Journal* 7: 106–116.

———. 2008. Mata matters: Macros. *Stata Journal* 8: 401–412.

Gould, W., J. Pitblado, and B. Poi. 2010. *Maximum Likelihood Estimation with Stata*. 4th ed. College Station, TX: Stata Press.

Greenland, S. 2000. Principles of multilevel modelling. *International Journal of Epidemiology* 29: 158–167.

Han, C., and B. P. Carlin. 2001. Markov chain Monte Carlo methods for computing Bayes factors: A comparative review. *Journal of the American Statistical Association* 96: 1122–1132.

Hastings, W. K. 1970. Monte Carlo sampling methods using Markov chains and their applications. *Biometrika* 57: 97–109.

Hilbe, J. M. 2011. *Negative Binomial Regression.* 2nd ed. Cambridge: Cambridge University Press.

Hitchcock, D. B. 2003. A history of the Metropolis–Hastings algorithm. *American Statistician* 57: 254–257.

Hoeting, J. A., D. Madigan, A. E. Raftery, and C. T. Volinsky. 1999. Bayesian model averaging: A tutorial. *Statistical Science* 14: 382–401.

Hogg, R. V., J. McKean, and A. T. Craig. 2013. *Introduction to Mathematical Statistics.* 7th ed. Upper Saddle River, NJ: Pearson.

Irony, T. Z., and N. D. Singpurwalla. 1997. Non-informative priors do not exist: A dialogue with José M. Bernardo. *Journal of Statistical Planning and Inference* 65: 159–177.

Johnson, S. R., G. A. Tomlinson, G. A. Hawker, J. T. Granton, and B. M. Feldman. 2010. Methods to elicit beliefs for Bayesian priors: A systematic review. *Journal of Clinical Epidemiology* 63: 355–369.

Kass, R. E., and A. E. Raftery. 1995. Bayes Factors. *Journal of the American Statistical Association* 90: 773–795.

Kerman, J., A. Gelman, T. Zheng, and Y. Ding. 2008. Visualization in Bayesian data analysis. In *Handbook of Data Visualization*, ed. C. Chen, W. Härdle, and A. Unwin, 709–724. Berlin: Springer.

Kirke, P. N., J. L. Mills, A. M. Molloy, L. C. Brody, V. B. O'Leary, L. Daly, S. Murray, M. Conley, P. D. Mayne, O. Smith, and J. M. Scott. 2004. Impact of the MTHFR C677T polymorphism on risk of neural tube defects: Case–control study. *British Medical Journal* 328: 1535.

Kyung, M., J. Gill, M. Ghosh, and G. Casella. 2010. Penalized regression, standard errors, and Bayesian lassos. *Bayesian Analysis* 5: 369–412.

Lawless, J. F. 1987. Negative binomial and mixed Poisson regression. *Canadian Journal of Statistics* 15: 209–225.

Lee, Y., J. A. Nelder, and Y. Pawitan. 2006. *Generalized Linear Models with Random Effects: Unified Analysis via H-likelihood.* Boca Raton, FL: Chapman & Hall/CRC.

Lunn, D. J., C. Jackson, N. G. Best, A. Thomas, and D. J. Spiegelhalter. 2013. *The BUGS book: A Practical Introduction to Bayesian Analysis.* Boca Raton, FL: CRC Press.

Lunn, D. J., A. Thomas, N. G. Best, and D. J. Spiegelhalter. 2000. WinBUGS – A Bayesian modelling framework: Concepts, structure, and extensibility. *Statistics and Computing* 10: 325–337.

Margolin, B. H., N. Kaplan, and E. Zeiger. 1981. Statistical analysis of the Ames *Salmonella*/microsome test. *Proceedings of the National Academy of Sciences* 78: 3779–3783.

Marshall, E. C., and D. J. Spiegelhalter. 2003. Approximate cross-validatory predictive checks in disease mapping models. *Statistics in Medicine* 22: 1649–1660.

Marske, D. 1967. *Biomedical oxygen demand data interpretation using sums of squares surface.* Master's thesis, University of Wisconsin–Madison.

Metropolis, N., A. W. Rosenbluth, M. N. Rosenbluth, A. H. Teller, and E. Teller. 1953. Equations of state calculations by fast computing machines. *Journal of Chemical Physics* 21: 1087–1092.

Morris, J. K., D. E. Mutton, and E. Alberman. 2002. Revised estimates of the maternal age specific live birth prevalence of Down's syndrome. *Journal of Medical Screening* 9: 2–6.

Natarajan, R., and C. E. McCulloch. 1998. Gibbs sampling with diffuse proper priors: A valid approach to data-driven inference? *Journal of Computational and Graphical Statistics* 7: 267–277.

Neal, R. M. 1998. Suppressing random walks in Markov chain Monte Carlo using ordered overrelaxation. *Learning in Graphical Models* 89: 205–228.

———. 2003. Slice sampling. *Annals of Statistics* 31: 705–767.

Ntzoufras, I. 2009. *Bayesian Modeling Using WinBUGS.* Hoboken, NJ: Wiley.

Nummelin, E. 2002. MC's for MCMC'ists. *International Statistical Review* 70: 215–240.

Oman, S. D., N. Meir, and N. Haim. 1999. Comparing two measures of creatinine clearance: An application of errors-in-variables and bootstrap techniques. *Journal of the Royal Statistical Society, Series C* 48: 39–52.

Park, T., and G. Casella. 2008. The Bayesian lasso. *Journal of the American Statistical Association* 103: 681–686.

Plummer, M. 2008. Penalized loss functions for Bayesian model comparison. *Biostatistics* 9: 523–539.

Plummer, M., N. G. Best, M. K. Cowles, and K. Vines. 2006. CODA: Convergence diagnosis and output analysis for MCMC. *R News* 6(1): 7–11.

Ratkowsky, D. A. 1983. *Nonlinear Regression Modeling: A Unified Practical Approach.* New York: Marcel Dekker.

Ritter, C., and M. A. Tanner. 1992. Facilitating the Gibbs sampler: The Gibbs stopper and the Griddy–Gibbs sampler. *Journal of the American Statistical Association* 87: 861–868.

Roberts, G. O., and J. S. Rosenthal. 2001. Optimal scaling for various Metropolis–Hastings algorithms. *Statistical Science* 16: 351–367.

Rubin, D. B. 1984. Bayesianly justifiable and relevant frequency calculations for the applied statistician. *Annals of Statistics* 12: 1151–1172.

Rutherford, M. J., P. C. Lambert, and J. R. Thompson. 2010. Age–period–cohort modeling. *Stata Journal* 10: 606–627.

Ryu, E. 2009. Simultaneous confidence intervals using ordinal effect measures for ordered categorical outcomes. *Statistics in Medicine* 28: 3179–3188.

Saha, K., and S. Paul. 2005. Bias-corrected maximum likelihood estimator of the negative binomial dispersion parameter. *Biometrics* 61: 179–185.

Schwarz, G. 1978. Estimating the dimension of a model. *Annals of Statistics* 6: 461–464.

Sinharay, S. 2004. Experiences with Markov chain Monte Carlo convergence assessment in two psychometric examples. *Journal of Educational and Behavioral Statistics* 29: 461–488.

Smith, B. J. 2007. boa: An R package for MCMC output convergence assessment and posterior inference. *Journal of Statistical Software* 21: 1–37.

Spiegelhalter, D. J. 1998. Bayesian graphical modelling: A case-study in monitoring health outcomes. *Journal of the Royal Statistical Society, Series C* 47: 115–133.

Spiegelhalter, D. J., N. G. Best, B. P. Carlin, and A. Van Der Linde. 2002. Bayesian measures of model complexity and fit. *Journal of the Royal Statistical Society, Series B* 64: 583–639.

Stiger, T. R., H. X. Barnhart, and J. M. Williamson. 1999. Testing proportionality in the proportional odds model fitted with GEE. *Statistics in Medicine* 18: 1419–1433.

Suess, E. A., and B. E. Trumbo. 2010. *Introduction to Probability Simulation and Gibbs Sampling with R*. New York: Springer.

Tanner, M. A. 1996. *Tools for Statistical Inference: Methods for the Exploration of Posterior Distributions and Likelihood Functions*. 3rd ed. New York: Springer.

Thompson, J. R., T. Palmer, and S. Moreno. 2006. Bayesian analysis in Stata with WinBUGS. *Stata Journal* 6: 530–549.

Tibshirani, R. 1996. Regression shrinkage and selection via the lasso. *Journal of the Royal Statistical Society, Series B* 58: 267–288.

Tierney, L. 1994. Markov chains for exploring posterior distributions. *Annals of Statistics* 22: 1701–1728.

Toft, N., G. T. Innocent, G. Gettinby, and S. W. J. Reid. 2007. Assessing the convergence of Markov chain Monte Carlo methods: An example from evaluation of diagnostic tests in absence of a gold standard. *Preventive Veterinary Medicine* 79: 244–256.

Wakefield, J. C., A. F. M. Smith, A. Racine-Poon, and A. E. Gelfand. 1994. Bayesian analysis of linear and non-linear population models by using the Gibbs sampler. *Journal of the Royal Statistical Society, Series C* 43: 201–221.

Warner, J. O. 2001. A double-blinded, randomized, placebo-controlled trial of cetirizine in preventing the onset of asthma in children with atopic dermatitis: 18 months' treatment and 18 months' posttreatment follow-up. *Journal of Allergy and Clinical Immunology* 108: 929–937.

Wright, D. E. 1986. A note on the construction of highest posterior density intervals. *Journal of the Royal Statistical Society, Series C* 35: 49–53.

Xie, T., and M. Aickin. 1997. A truncated Poisson regression model with applications to occurrence of adenomatous polyps. *Statistics in Medicine* 16: 1845–1857.

Yu, B., and P. Mykland. 1998. Looking at Markov samplers through cusum path plots: A simple diagnostic idea. *Statistics and Computing* 8: 275–286.

Author index

O

P

R

S

T

V

W

X

Y

Z

Subject index

G

H

I

J

L

M

N

O